COUNTERING AGRIC
BIOTERRORISM

Committee on Biological Threats to
Agricultural Plants and Animals

Board on Agriculture and Natural Resources
Board on Life Sciences

Division on Earth and Life Studies

NATIONAL RESEARCH COUNCIL
OF THE NATIONAL ACADEMIES

THE NATIONAL ACADEMIES PRESS
Washington, D.C.
www.nap.edu

THE NATIONAL ACADEMIES PRESS 500 Fifth Street, N.W. Washington, DC 20001

NOTICE: The project that is the subject of this report was approved by the Governing Board of the National Research Council, whose members are drawn from the councils of the National Academy of Sciences, the National Academy of Engineering, and the Institute of Medicine. The members of the committee responsible for the report were chosen for their special competences and with regard for appropriate balance.

This study was supported by Grant No. 59-0790-0-170 between the National Academies and the United States Department of Agriculture. Any opinions, findings, conclusions, or recommendations expressed in this publication are those of the author(s) and do not necessarily reflect the views of the organizations or agencies that provided support for the project.

International Standard Book Number 0-309-08545-4 (Book)
International Standard Book Number 0-309-50989-0 (PDF)
Library of Congress Catalog Card Number 2003104243

Additional copies of this report are available from the National Academies Press, 500 Fifth Street, N.W., Lockbox 285, Washington, DC 20055; (800) 624-6242 or (202) 334-3313 (in the Washington metropolitan area); Internet, http://www.nap.edu

Printed in the United States of America

THE NATIONAL ACADEMIES

Advisers to the Nation on Science, Engineering, and Medicine

The **National Academy of Sciences** is a private, nonprofit, self-perpetuating society of distinguished scholars engaged in scientific and engineering research, dedicated to the furtherance of science and technology and to their use for the general welfare. Upon the authority of the charter granted to it by the Congress in 1863, the Academy has a mandate that requires it to advise the federal government on scientific and technical matters. Dr. Bruce M. Alberts is president of the National Academy of Sciences.

The **National Academy of Engineering** was established in 1964, under the charter of the National Academy of Sciences, as a parallel organization of outstanding engineers. It is autonomous in its administration and in the selection of its members, sharing with the National Academy of Sciences the responsibility for advising the federal government. The National Academy of Engineering also sponsors engineering programs aimed at meeting national needs, encourages education and research, and recognizes the superior achievements of engineers. Dr. Wm. A. Wulf is president of the National Academy of Engineering.

The **Institute of Medicine** was established in 1970 by the National Academy of Sciences to secure the services of eminent members of appropriate professions in the examination of policy matters pertaining to the health of the public. The Institute acts under the responsibility given to the National Academy of Sciences by its congressional charter to be an adviser to the federal government and, upon its own initiative, to identify issues of medical care, research, and education. Dr. Harvey V. Fineberg is president of the Institute of Medicine.

The **National Research Council** was organized by the National Academy of Sciences in 1916 to associate the broad community of science and technology with the Academy's purposes of furthering knowledge and advising the federal government. Functioning in accordance with general policies determined by the Academy, the Council has become the principal operating agency of both the National Academy of Sciences and the National Academy of Engineering in providing services to the government, the public, and the scientific and engineering communities. The Council is administered jointly by both Academies and the Institute of Medicine. Dr. Bruce M. Alberts and Dr. Wm. A. Wulf are chair and vice chair, respectively, of the National Research Council.

www.national-academies.org

COMMITTEE ON BIOLOGICAL THREATS TO AGRICULTURAL PLANTS AND ANIMALS

HARLEY W. MOON (*Chair*), Veterinary Medical Research Institute, Iowa State University, Ames, Iowa
MICHAEL ASCHER, California Department of Health Services, Richmond, California, and Office of Public Health Preparedness, Department of Health and Human Services, Washington, DC
R. JAMES COOK, Washington State University, Pullman, Washington
DAVID R. FRANZ, Southern Research Institute, Frederick, Maryland
MARJORIE HOY, University of Florida, Gainesville, Florida
DONALD F. HUSNIK, Consultant, US Department of Agriculture Animal and Plant Health Inspection Service (retired)
HELEN H. JENSEN, Iowa State University, Ames, Iowa
KENNETH H. KELLER, Hubert H. Humphrey Institute of Public Affairs, University of Minnesota, Minneapolis, Minnesota
JOSHUA LEDERBERG, The Rockefeller University, New York, New York
LAURENCE V. MADDEN, The Ohio State University, Wooster, Ohio
LINDA S. POWERS, National Center for the Design of Molecular Function, Utah State University, Logan, Utah
ALFRED D. STEINBERG,[1] The MITRE Corporation, McLean, Virginia

Consultant

AL STRATING, Consultant, Animal Health Solutions International, US Department of Agriculture Animal and Plant Health Inspection Service (retired)

Liaison, Committee on Agricultural Biotechnology, Health, and the Environment

ROBERT E. SMITH, President, R.E. Smith Consulting, Inc., Newport, Vermont

Staff

JENNIFER KUZMA, Study Director
NORMAN GROSSBLATT, Editor
LAURA HOLLIDAY,[2] Research Assistant
DEREK SWEATT,[3] Research Assistant
SETH STRONGIN,[4] Project Assistant

---

[1]March 2001–January 2002.
[2]August 2001–April 2002.
[3]Through August 2001.
[4]Since March 2002.

BOARD ON LIFE SCIENCES

COREY S. GOODMAN (*Chair*), University of California, Berkeley, California
R. ALTA CHARO, University of Wisconsin, Madison, Wisconsin
JOANNE CHORY, The Salk Institute for Biological Studies, La Jolla, California
DAVID J. GALAS, Keck Graduate Institute of Applied Life Science, Claremont, California
BARBARA GASTEL, Texas A&M University, College Station, Texas
JAMES M. GENTILE, Hope College, Holland, Michigan
LINDA E. GREER, Natural Resources Defense Council, Washington, DC
ED HARLOW, Harvard Medical School, Boston, Massachusetts
ELLIOT M. MEYEROWITZ, California Institute of Technology, Pasadena, California
ROBERT T. PAINE, University of Washington, Seattle, Washington
GREGORY A. PETSKO, Brandeis University, Waltham, Massachusetts
STUART L. PIMM, Columbia University, New York, New York
JOAN B. ROSE, University of South Florida, St. Petersburg, Florida
GERALD M. RUBIN, Howard Hughes Medical Institute, Chevy Chase, Maryland
BARBARA A. SCHAAL, Washington University, St. Louis, Missouri
RAYMOND L. WHITE, DNA Sciences, Incorporated, Fremont, California

Staff

FRANCES E. SHARPLES, Director
BRIDGET K.B. AVILA, Senior Project Assistant

Foreword

The National Academies are releasing this report, *Countering Agricultural Bioterrorism*, in the midst of a vigorous national debate over what scientific information relevant to countering terrorism should (and should not) be made public. The scientific community is deeply engaged in this debate over the issue of "sensitive but unclassified information," seeking to find the appropriate balance between the open interchange of ideas and findings that is required to advance scientific understanding and meet human needs, versus a desire to prevent information from empowering individuals and groups to do harm. In this case, we are convinced that this report will increase our security by helping to inform and assist the nation in improving its awareness, capabilities, and plans to defend against threats of agricultural bioterrorism. We also believe that it can usefully inform the current deliberations on how best to organize the federal government to address homeland security.

The National Academies convened a diverse committee of twelve experts in 2000 in response to a request from, and with the financial support of, the US Department of Agriculture (USDA). This committee was charged with the task of evaluating US preparedness for biological threats to agricultural plants and animals, and was asked to recommend ways to improve the ability of federal agencies to protect the nation's supply of food and fiber. The committee was well into its deliberations before the United States was attacked on September 11, 2001, and it had identified many strengths and weaknesses of the US system by the end of that year. Because of the importance of this study to homeland defense, the National Academies made a special exception to their normal practice of maintaining strict confidentiality of studies until public release by authorizing a briefing in March 2002 of White House, homeland security, and agricultural federal officials on the committee's findings and preliminary conclusions. When a draft of the full report was completed, it was submitted to the Department of Agriculture and subsequently the Office of Homeland Security for classification review. Because the Government has been aware of the findings and conclusions of the committee for some time, it is likely that the government has already taken actions to respond to at least some of them. The report should therefore be read with this qualification in mind.

Consistent with our Congressional charter, the National Academies have provided independent scientific, engineering, and medical advice to the Federal Government since 1863 in published reports. It has always been the policy of the National Academies that all reports, even those produced with access to national security classified information, be publicly disseminated in an unclassified version. Since the 1997 enactment of Section 15 of the Federal Advisory Committee Act (FACA), the National Academies are legally required to do so. In their public reports, the Academies are permitted under Section 15 to protect from public release only information that is classified or satisfies one of the other exemptions to the Freedom of Information Act.

After discussions between leaders at the National Academies and officials at the USDA and the Office of Homeland Security over the past several months, USDA asked the National Academies to withhold the full report from the public. However, the government did not assert that the report contained classified information, and the National Academies have concluded that only limited portions of the report, described below, could fall under the conditions of Section 15 of FACA for withholding information. Therefore, we have decided to issue this public version of *Countering Agricultural Bioterrorism* at this time.

In considering a written request from USDA, and consistent with FACA, the National Academies have removed detailed and specific information that was originally part of Chapter 3 of the report, as well as certain other portions of the text of the report that contain detailed and specific discussions of certain matters. The original Chapter 3 has been produced as an appendix that is not for distribution to the general public.

This report on agricultural bioterrorism is but one of many examples of the National Academies' efforts to help the nation increase its security. The recently released 440 page report, entitled *Making the Nation Safer: The Role of Science and Technology in Countering Terrorism,* which outlines how the Federal Government can use science to improve homeland security, is another—as is *Responding First to Bioterrorism*, a new National Academies website (*www.nap.edu/shelves/first/*) that gives police, firefighters, and public safety officials access to the latest science and methods for responding to a biological terror attack.

The events of September 11, 2001 have underscored the threat of terrorism that grips the nation and world. Science and scientists can reduce the threat. The National Academies are committed to helping them to do so. We wish to thank the committee members who prepared this report, as well as the staff, consultants, and reviewers who assisted them. We believe that their efforts have produced an important contribution to our future welfare.

Bruce Alberts
President
National Academy of Sciences

William A. Wulf
President
National Academy of Engineering

Harvey V. Fineberg
President
Institute of Medicine

Preface

Concerns about the potential public health and economic consequences of agricultural bioterrorist attacks against the United States were broadly identified during a National Research Council (NRC) workshop in August 1999. The United States Department of Agriculture's (USDA) Agricultural Research Service (ARS) subsequently commissioned the NRC to conduct the study of intentional biological threats to agricultural plant and animals, reported herein. The study was a collaborative effort of the NRC Board on Agriculture and Natural Resources and the Board on Life Sciences. Brief biographies of the committee of volunteers who conducted the study and those who facilitated the work in consultant, staff, and liaison capacities are listed in Appendix E. The study committee had expertise in molecular biology, microbiology, entomology, plant pathology, veterinary medicine, public health, biological warfare, chemical engineering, regulatory affairs, agricultural economics, and public policy.

The formal statement of task for the committee is provided below:

> In order to evaluate US preparedness for biological threats directed towards agricultural plants and animals, the US Department of Agriculture asked the NRC Board on Agriculture and Natural Resources to convene a diverse group of experts to evaluate the ability of the United States to deter, prevent, detect, thwart, respond to and recover from an intentional, biological attack against the nation through its supply of food and fiber. This NRC study will emphasize deterrence and prevention and will include an analysis of the defensive posture before an attack takes place as well as capability to respond to an attack. The study will focus on two key areas and how these two areas are intertwined:

> **1) US System.** A selected set of biological agents, representative of various types of potential threats, will be used to examine, under different scenarios, the probable response and efficacy of the current agricultural protection system and related health, law enforcement and intelligence systems. The study will identify the key organizations and participants in these systems at federal, state, and local levels, and will evaluate their respective roles and critical interactions to defend against an attack in the first place and to respond in the event of a biological attack. These organizations will include public health agencies, the intelligence community, law enforcement and other agencies that will need to partner with the US agricultural system. Short-term (e.g., government responsiveness) and long-term (e.g., research and education) national defense strategies will be evaluated.
>
> **2) Use of Science in the US System.** The study committee will review the use of scientific information in providing the basis of defense policy and procedures to defend US agriculture, and will explore the potential of microbial forensics, vaccine/drug development, biochemical detection, decontamination technology, molecular epidemiology, and other technologies as tools to deter, prevent, thwart or resist a biological threat. Scientific research needed to enhance agricultural defense capabilities will be identified where appropriate.
>
> The study committee will make recommendations, if necessary, on ways to improve the defensive posture, national security and scientific capabilities of the US agricultural and food system and the preparedness of federal agencies to catalyze action, cooperate, and effectively respond in order to protect the nation's supply of food and fiber.

The committee met five times to review the US system for countering agricultural bioterrorism, identify the strengths and weaknesses of that system, and come to consensus on conclusions and recommendations. One of the meetings included a public workshop with dialogue among committee members and invited stakeholders from a broad array of public and private sector interests relevant to the study. In addition to this workshop, the committee had the benefit of perspectives presented by our sponsor as well as by experts on the intelligence, global laboratory surveillance, military and social–psychological implications of agricultural bioterrorism at other public sessions during our meetings. Information from the workshop and public sessions contributed greatly to the committee's understanding of the US defense system as well as the use of science and technology within that system. On behalf of the committee, I thank the workshop and other public session participants listed in Appendix B for their contributions to our work.

Each committee member reviewed, commented on (and where appropriate, debated, and revised) and ultimately accepted the entire report. The report presents the committee's consensus understanding, analysis, judgment and recommendations. The draft consensus report was reviewed external to NRC by individuals with expertise relevant to diverse areas addressed in the report. Such external review is NRC Report Review Committee policy to assist in delivery of an objective, evidence-driven, sound, balanced final report that meets the study charge. Reviewers remained unknown to the study committee until the final report was released. The study committee considered all suggestions and comments from the reviewers, and where appropriate, modified the draft into the final report. On behalf of the study committee, I thank the reviewers (listed in Acknowledgments) for their assistance in developing our final report. The report is the responsibility of the NRC and the study committee, not that of the reviewers, who did not see the final report until it was published.

Chapter 1 introduces the report with a brief overview of the historical, social and political, economic and public policy dimensions of agricultural bioterrorism. It summarizes the charge to the committee and defines some terms as used in the report. Chapter 2 presents an overview and identifies strengths and weaknesses of the US system for countering agricultural bioterrorism. It points out that the system was designed to protect against unintentional threats, and as such is useful, but inadequate, for protection against the intentional threats of terrorists. Chapter 3 presents lessons learned from the committee's in-depth deliberations and analyses of various potential threat agents. Chapter 4 presents research needs and emerging science and technology that provides major opportunities for strengthening our defense if that knowledge and technology is developed for and applied to agriculture.

Chapter 5 contains the committee's principal findings, conclusions, recommendations, and supporting text explaining our rationale. It summarizes the nation's vulnerabilities and presents a framework for a comprehensive national plan to counter them. The plan recognizes the paramount importance of early detection and response at the local and state levels, as well as the need for federal leadership in developing and implementing the plan.

On behalf of the committee I would like to thank our study director, Dr. Jennifer Kuzma, for her commitment, probing questions, and able direction throughout the study. We are also grateful for the support and coordination provided by our research assistants, Derek Sweatt and Laura Holliday, and our project assistant, Seth Strongin. The work of these four NRC staff members was essential to the timely conclusion and quality of the report. I would also like to personally thank the members of the committee for the objectivity, energy and collegiality with which they approached the study. It has been a pleasure as well as a privilege to work with them.

Harley W. Moon
Chair
Committee on Biological Threats to
Agricultural Plants and Animals

Acknowledgments

Many individuals contributed to this study, and it would not have been possible without their dedication of time and expertise. Specifically, the committee would like to thank the individuals listed in Appendix A, who participated in the 1999 National Research Council (NRC) planning workshop, which set the groundwork for this study. Likewise, during the course of this study, several stakeholders and experts freely devoted their time to share their thoughts and ideas in open sessions with the committee. Those individuals are listed in Appendix B. Their insights were invaluable to this report. In addition, the committee would like to thank Dr. Joan Arnoldi for her contributions as a committee member during the first two months of the study.

The committee would also like to thank the oversight boards of the NRC and their staff, as listed in the preceding pages. Their commitment to this effort is greatly appreciated.

This report has been reviewed in draft form by individuals chosen for their diverse perspectives and technical expertise, in accordance with procedures approved by the NRC's Report Review Committee. The purpose of this independent review is to provide candid and critical comments that will assist the NRC in making its published report as sound as possible and to ensure that the report meets institutional standards for objectivity, evidence, and responsiveness to the study charge. The review comments and draft manuscript remain confidential to protect the integrity of the deliberative process. We thank the following individuals for their review of this report:

Steven M. Becker, The University of Alabama at Birmingham
May Berenbaum, University of Illinois

Ron Carroll, University of Georgia
Bob Hillman, Idaho State Department of Agriculture
Col. Nancy Jaxx, US Medical Research Institute of Infectious Diseases
Jan Leach, Kansas State University
Whitney MacMillan, Cargill, Incorporated
Randy Murch, Federal Bureau of Investigation
Gil Omenn, University of Michigan
Craig Reed, Virginia Tech College of Veterinary Medicine
Luis Sequeira, University of Wisconsin
Isi Siddiqui, Crop Life America
Daniel Sumner, University of California, Davis
David Walt, Tufts University
George Whitesides, Harvard University
Major Neal Woollen, US Medical Research Institute of Infectious Diseases

Although the reviewers above have provided many constructive comments and suggestions, they were not asked to endorse the conclusions or recommendations nor did they see the final draft of the report before its release. The review of this report was overseen by John Bailar, University of Chicago, and Dale Bauman, Cornell University. Appointed by the NRC, they were responsible for making certain that an independent examination of this report was carried out in accordance with institutional procedures and that all review comments were carefully considered. Responsibility for the final content of this report rests entirely with the authoring committee and the NRC.

Contents

TABLES, FIGURES, AND BOXES

Tables

Figures

Boxes

Executive Summary

Public confidence in the security of the US food and fiber system has been sustained by the quality, variety, abundance, and affordability of agricultural products in the United States. Although the system in place to defend against unintentional threats to agriculture has weaknesses and needs, the demonstrated ability of the system to resolve, accommodate, or manage critical food safety problems, temporary shortages of some commodities, plant and animal infestations and diseases, and natural disasters indicates that, in general, such confidence has been warranted. However, over the last several years, there has been recognition of the possibility and consequences of intentional threats directed at US agriculture. Such attacks could come from foreign or domestic terrorists and use biological, chemical, or radiological agents. They could be directed at the pre harvest (live plant and live animal) or post harvest (processing and distribution) stages of food and fiber production.

Historically, incidents have involved biological attacks directed at plants and animals, and offensive programs in several countries have been directed against agriculture. In addition, on September 11, 2001, the most detrimental and shocking terrorist attacks on US soil occurred. Since then, the nation has changed its perspective on the likelihood of terrorism and its vulnerability to it. Also during the fall of 2001, for the first time in history, a sophisticated preparation of a biological weapons agent, *Bacillus anthracis* (anthrax), was successfully used to cause death, disruption, and terror. That provided concrete evidence of the desire and ability on the part of some to use biological weapons on US soil. Bioterrorism is now a reality.

SCOPE AND APPROACH OF THE STUDY

In light of the above, there is a need to assess the vulnerability of US agriculture to intentional threats. The National Research Council (NRC) started the process in August 1999 by hosting a planning workshop on agricultural bioterrorism (Appendix A). The workshop identified a need for a study of agricultural bioterrorism and served as a catalyst for the present study. This study of intentional biological threats to agricultural plants and animals was requested by the US Department of Agriculture (USDA) Agricultural Research Service (ARS). The study committee met five times during the period May 2001-January 2002. Several of the meetings included public sessions to obtain input from diverse experts and stakeholders (Appendix B).

At the time the study was initiated in early 2001, concerns about possible terrorist attacks on US agriculture were not prevalent. Terrorist attacks during the fall of 2001 dramatically changed the prevailing attitude. Economic terrorism in the form of biological attacks on agriculture is now widely perceived as a threat to the nation. There is general recognition of the need to defend the nation against such threats, and additional resources have been allocated to the USDA and other departments to do so.

The charge to this committee was to evaluate the ability of the United States to deter, prevent, detect, thwart, respond to, and recover from intentional biological attacks on the nation at the live plant and live animal stage of food and fiber production (see Preface and Box 1-2 for the full text of the committee's statement of task). The committee's charge did not include issues associated with the post harvest contamination of the food supply, although some of these issues are discussed throughout the text. Likewise, the committee was not asked to develop a priority-order threat list; however, the committee conducted in-depth study of plausible threat and target combinations to examine the strengths and weaknesses of the US system.

This summary and Chapter 5 contain the committee's general findings, conclusions and recommendations. Chapters 2, 3, and 4 support them.[1] Other, more specific findings, conclusions, and recommendations are presented in bold face at the end of Chapter 4 and general lessons learned are presented in Chapter 3.

This report includes a profile of the US system for defense against biological threats to agriculture. The bulk of the committee's analysis took place in late 2001 (approximately August 2001 to December 2001), but as new information was received, the committee reexamined and updated its analysis until spring

[1]In chapter 5, supporting sections are referenced. Likewise, in supporting sections in the body of the report, references to the numbered findings, conclusions, and recommendations in chapter 5 are made.

2002.[2] It identifies changes needed to strengthen and adapt the system for protection against intentional biological threats. Actions to address some of the needs identified in this report have been proposed and are being debated by the US government and agricultural community or are in the early stages of implementation (Chapter 2 and Appendix D). The committee hopes that this report will help to inform the debate and the implementation of change.

KEY FINDINGS AND CONCLUSIONS

The committee's main findings and conclusions are summarized in Box ES-1. Overall, the committee found that US agriculture is indeed vulnerable to bioterrorism. The findings and conclusions that support this overarching view appear in Box ES-1 and are discussed below.

Finding I

Biological threat agents that vary in the nature and extent of their potential impact on plant crops and livestock are widely available for intentional introduction and pose a major threat to US agriculture. Technical sophistication would not be necessary for attacks with some of these threat agents. Although an attack with such agents is highly unlikely to result in famine or malnutrition, the possible damage includes major direct and indirect costs to the agricultural and national economy, adverse public-health effects (especially if the agents are zoonotic), loss of public confidence in the food system and in public officials, and widespread public concern and confusion.

Finding I.A

There are weaknesses in the US agricultural defense against unintentional biological threats. However, it is imperative to recognize that without specific planning, research, and programs for intentional threats, merely fixing those weaknesses will not adequately protect the United States against *intentional* attacks for several reasons. The perpetrators will have the advantage of selecting unanticipated and covert means, including control of the time of introduction of the agent, introducing agents into remote areas, multiple introductions of the same agent, and simultaneous release of different agents. Intentional introductions permit an increased probability of survival of a pest or pathogen in transit, the use of

[2]The proposal to establish a new Department of Homeland Security, and transfer authority for the United States Department of Agriculture's Animal and Plant Health Inspection Service to it, was announced by President Bush on June 18, 2002. This announcement occurred after external review of the report, and therefore, the committee did not incorporate an analysis of it.

BOX ES-1
Findings and Conclusions

I. The United States is vulnerable to bioterrorism directed against agriculture.

   A. Intentional introductions of pests and pathogens may differ substantially from unintentional introductions.
   B. The nation has inadequate plans to deal with agricultural bioterrorism.

      1. As of spring 2002, no publicly available, in-depth, interagency or interdepartmental national plan had been formulated for defense against the intentional introduction of biological agents directed at agriculture.
      2. The adverse effects of bioterrorism agents on wildlife have been little considered.

   C. There are important gaps in our knowledge of foreign-plant and foreign-animal pests and pathogens. These gaps reduce the reliability and timeliness of risk assessments and risk-management decisions.
   D. The current inspection and exclusion program at the US borders, in which only small proportions of people and goods entering the United States are inspected, is inadequate for countering the threat of agricultural bioterrorism.
   E. Our ability to rapidly detect and identify most plant pests and pathogens and some animal pests and pathogens soon after introduction is inadequate. This allows them to spread, results in greater damage, and makes it more expensive or impossible to respond with eradication.
   F. A large-scale multifocal attack on agriculture could not be responded to or controlled adequately or quickly and would overwhelm existing laboratory and field resources.
   G. It is not feasible to be specifically prepared or have all the scientific tools for every contingency or threat to agriculture.
   H. Although the nation's fundamental science, research, and education infrastructure (academic, industrial, and government) is in place and functional, preparing the nation for agricultural bioterrorism requires special efforts and support of the infrastructure.
   I. There is a need to enhance the basic understanding of threat agents so as to develop new and exploit emerging technologies for rapid detection, identification, prophylaxis, and control.

highly virulent strains and high concentrations of inoculum, and precise timing of release to coincide with maximal colonization potential. These attacks also differ from unintentional introductions in that the perpetrator is able to target susceptible production areas and natural environments, while sabotaging laboratory and field-testing resources.

Finding I.B

The USDA Animal and Plant Health Inspection Service (APHIS) has several emergency plans for dealing with unintentional introductions of plant and animal pests and pathogens, but as of spring 2002, the committee could not find any publicly-available interdepartmental national plan designed for defense against an intentional introduction of a plant or animal pest or pathogen.[3] Likewise, there is no accepted interagency threat list for agricultural bioterrorism that can be used in developing a defense response plan that includes coordination among partners, laboratory strategy, public information, and determination of research needs.[4] Coordination within and among key federal agencies, as well as coordination of federal agencies with state and local agencies and private industry, appears to be insufficient for effectively deterring, preventing, detecting, responding to, and recovering from agricultural threats. A well-organized plan is crucial for mitigating the adverse impacts of such attacks.

Finding I.C

The basic biology of many agricultural pests and pathogens is not well understood. In addition, improved surveillance is needed. Threat assessments need to be based on global intelligence concerning hostile government and terrorist activities and on continuing international surveillance for new and emerging diseases and infestations, and they need to be regularly updated. Assessments (conducted jointly by agencies responsible for intelligence and agriculture) of the most probable means of acquiring, introducing, and disseminating each of the major intentional threat agents are needed.

Finding I.D

Current methods for interdicting pests and pathogens at borders are not targeted at intentional biological threats. In addition, only a small percentage of travelers or imports are actually inspected by APHIS (for example, 1% of pri-

[3]The committee recognizes that such plans might be classified and could not be openly discussed with the committee. The committee has received no indication, however, that such plans exist.

[4]The committee recognizes that it might be in the best interest of national security to keep this list classified or confidential (i.e., not available to the general public or potential terrorists).

vately owned vehicles). New technologies and novel inspection methods are needed that improve our ability to detect key pests and pathogens at seaports and border crossing sites. Faster, more sensitive, and more cost-effective exclusion methods are needed.

Finding I.E

Border controls are an integral component of a multitactic strategy to reduce the threat of agricultural bioterrorism, and they reduce the risks somewhat. However, they cannot be relied on to prevent intentional introduction. Sophisticated terrorists are likely to devise ways to circumvent such controls. Thus, a capacity for rapid detection and response after introduction is of paramount importance in countering agricultural bioterrorism.

Time is of the essence in minimizing damage in the event of an agricultural bioterrorism attack. The lag between pest or pathogen establishment and detection directly affects our ability to respond adequately. Currently, no comprehensive system is in place for screening, identifying, and reporting pests or pathogens found by passive surveillance (informal surveillance through growers, extension agents, and so on). Likewise, the ability to detect and identify agricultural pests and pathogens varies from state to state.

Finding I.F

Current laboratory and field resources are often strained by naturally occurring outbreaks. For example, university plant clinics are typically understaffed and lack resources. Given the limited personnel, there is seldom time to perform molecular or biochemical assays on routine samples. In the event of a multifocal attack on agriculture, resources would not be sufficient for response or control. A surge capacity,[5] or agricultural bioterrorism rapid-response strategy, that can draw on industry, state and academic resources in times of crisis is needed.

Finding I.G

Although recent legislation has provided additional funding for combating agricultural bioterrorism, resources will always be limited compared to the large number of possible scenarios. Therefore, it is only feasible for plans to address a subset of the possible threats. This subset needs to cover diverse agent and target species that seem plausible to biologists, agricultural specialists, law enforce-

[5]Surge capacity or response is defined as a program, plan or capability that is designed to exceed normal capabilities in the event of an emergency situation. An analogous system would be the federal army reserves, which are asked to help in times of crisis, but are not always deployed.

ment, and the intelligence community. If effectively used for planning, this subset can also help to prepare for other possible scenarios.

Findings I.H and I.I

Information on and strategies aimed at protection against unintentionally introduced pests and pathogens will help but are insufficient for protection against deliberate introductions (**Finding I.A**). Future scientific research and programs need to encompass the unique challenges presented by deliberate introductions. The ability to prevent, detect, respond, and recover from an attack depends on the availability of sound information on the biology of the introduced organisms, the susceptibility of agricultural plants and animals to the threats, and the pathogenesis of the resulting diseases. From this perspective, large gaps exist in the fundamental knowledge of even the most common agricultural pests and pathogens and their effects.

KEY RECOMMENDATION: A COMPREHENSIVE PLAN

On the basis of the findings and conclusions that 1) the United States is vulnerable to agricultural bioterrorism and 2) insufficient plans are in place to defend against it, the committee recommends that the US government develop a comprehensive plan to counter agricultural bioterrorism (Box ES-2). This plan should include an implementation strategy that assigns responsibilities to agencies and organizations and holds them accountable for its development.

Elements of this plan are summarized in the text below, which refers to the numbered recommendations in Box ES-2.

The committee recommends a plan that will require significant additional resources for the federal, state, and local agencies and organizations involved in its implementation. The committee recognizes that decisions to direct resources to countering agricultural bioterrorism will need to be weighed against decisions to direct resources to other societal problems, including other forms of terrorism. However, it was beyond the committee's scope to perform cost-benefit analyses and pass judgment on the relative importance of agricultural bioterrorism.

Recommendation I.A

The committee reviewed various points in the food and fiber system where intervention to deter, prevent, thwart, detect, respond, and recover can take place. It is unlikely that intervention at any one point can provide acceptable security against intentional introductions (for example, at US borders). A series of interventions is more likely to reduce the threat of introduction or mitigate consequences after introduction. Therefore, multiple elements for countering agricultural bioterrorism should be incorporated into the plan. Also, the committee

BOX ES-2
Recommendations for a Comprehensive Plan to Counter Agricultural Bioterrorism

I. The US government should establish a comprehensive plan to respond to the threat of agricultural bioterrorism that does the following:

A. Integrates elements of deterrence, prevention, detection, response, and recovery.
B. Includes domestic and international strategies for recognition, prevention, and control.
C. Defines legal and jurisdictional authority and lead roles at the federal, state, local, and private levels and includes specifications for interagency cooperation.
D. Defines a categorical priority list of threat agents for planning.
E. Establishes operational capacity in:
 1. Surveillance, laboratory diagnosis, and electronic reporting of threat agents with
  a. Domestic and international surveillance for selected agricultural bioterrorism agents.
  b. A laboratory-response network for the detection, identification, and specific diagnosis of pest infestations and plant and animal diseases that might result from agricultural bioterrorism.
  c. Accelerated evaluation, validation, and adoption of emerging technologies for rapid detection and identification of threat agents, including agents modified by recombinant-DNA methods.
  d. A nationwide system (for example, analogous to the Centers for Disease Control and Prevention's Health Alert Network) for communication, data management and analysis, information dissemination, and education.
 2. Response to, cleanup of, and recovery from a bioterrorism event with
  a. Development and exercising of model scenarios of bioterrorism attacks using pathway analysis.
  b. Development of appropriate eradication and management plans in advance of attack for selected agricultural pests and pathogens, including chemotherapeutics, vaccination, or plant- or animal-breeding programs.
  c. Coordination among agricultural, wildlife, public health, human services, emergency management, intelligence, and law enforcement programs.
  d. International cooperation and outreach to assist other countries in managing or eradicating agricultural threat agents.

continues

3. Public information, education and outreach with
    a. Establishment and training of credible spokespersons for classes of threat agents.
    b. Development of specific media and public information for threat agents and outbreaks, including Internet-based information and training programs.
    c. Training of local officials in mass-media and public-information responses.
    d. A comprehensive educational program for the agricultural community at large to increase recognition of infestation and disease and improve response.

F. Establishes a comprehensive science and technology base to:
1. Increase basic understanding of
    a. Biology and epidemiology (including molecular epidemiology) of exotic agents in native and new environments.
    b. Pathogenesis in animals and plants (including through microbial genomics).
    c. Social and psychological impacts of agricultural terrorism.
    d. Perpetrators of terrorism, as one approach to deter, prevent, or thwart bioterrorist attacks.
2. Develop
    a. Interdiction, detection, diagnostic, and identification tools.
    b. Bioinformatics and information technology.
    c. Prophylaxis and therapeutics.
    d. Control and eradication strategies and tools.

G. Establishes a public-private advisory council on agricultural bioterrorism at the level of the Secretary of the US Department of Agriculture.

recognizes that a system for defense against unintentional threats is in place and recommends building upon that existing infrastructure (for example, APHIS emergency response) with special planning, research, and programs for intentional threats.

Recommendation I.B

The committee reviewed global and domestic strategies for recognizing potential biological threats to plants and animals, preventing attacks with these agents, and ultimately controlling them. Although there are some efforts to prepare for unintentional threats, there are no concerted national or international strategies for defense against intentional introductions of plant and animal pests or pathogens. Current procedures for defense against threats rely heavily on the voluntary action of the involved parties. Therefore, the committee recommends that domestic and international strategies be developed for recognition, prevention, and control of intentionally introduced plant and animal pests and pathogens.

Recommendation I.C

A federal plan for unintentional introductions of animal disease exists and is summarized in this report. However, as of spring 2002, there was no plan for intentional introductions of plant and animal threats (**Finding I.B.1**). It is unclear how the federal agencies would cooperate with each other and with other sectors during an attack. Therefore, the committee recommends that the overall plan for countering bioterrorism include delineation of roles and coordination of USDA with key partners—for example, the Federal Emergency Management Agency, defense, intelligence, public health, law enforcement, state and local agencies, private industry, trade associations, universities, human services organizations, and key global partners.

Recommendation I.D

The theoretical list of intentional biological threats to agriculture is broad, and it is not feasible to be specifically prepared or have all the scientific tools for every possible threat (**Finding I.G**). Although several lists of biological threats to agricultural plants and animals exist, they have been based on different sets of criteria. The current system needs to be evaluated and redesigned to respond appropriately to intentional threats. Therefore, the committee recommends that a short, interagency threat list be developed and specific planning conducted for each agent on the list to determine proper roles, resource allocations, laboratory strategy, research priorities, and so on.

Recommendation I.E

Information concerning agricultural diseases and infestations in foreign countries is often sketchy and not always timely. Furthermore, little information on biological-weapons R&D is available to the intelligence communities (**Finding I.C**). Therefore, the committee recommends studies to define the prevalence and epidemiology of major strains and subtypes of threat agents on a global basis. These studies can help predict where agents are most readily available and trace outbreaks to their origins after an attack. Through enhanced international surveillance, experts in all countries can be informed of each other's work and be more likely to notice shifts in the prevalence of biological agents or in R&D

designed to damage agriculture. The committee also recommends bolstering domestic surveillance.

Several steps need to be coordinated to detect and identify biological agents accurately and reliably. Laboratories need to work together to recognize threats at the earliest possible time and track the spread of biological agents on geographic and temporal scales. Otherwise, outbreaks might not be noticed for long periods. Time is critical for limiting the impact of an attack, recognizing that the event was intentional, and identifying the terrorist. Without appropriate laboratory capabilities, the severity or extent of a threat can be misjudged, and this can lead to greater agricultural, social, and human-health impacts.

Communication and coordination among laboratories are vital for determining appropriate and rapid response measures. There is no coordinated national system for laboratory analysis of biological threats directed at US agriculture. In the event of a large-scale or multifocal attack—or one using agents that persist in the environment or spread rapidly—existing laboratory systems and resources would be insufficient for supporting the response to minimize the spread and impact. The United States needs a laboratory network that can rapidly detect, identify, diagnose, report, and respond to agricultural bioterrorism threats. In this report, the committee discusses an existing laboratory network for human-health threats. Given the experience with the system and its proven ability to function in a crisis, the committee recommends that this framework be used as a starting point for developing a laboratory-response network for agriculture.[6]

The laboratory-response network should support tasks from field testing to biological forensics and build on existing federal, commercial, university, state extension, and regional facilities. It must be capable of performing molecular analyses, be continuously upgraded, and be connected by an electronic laboratory reporting system. The network should be used to identify laboratory capabilities and expertise. The network should also provide a surge capacity. Laboratories for agricultural, wildlife, and public-health diagnostics, as well as the Federal Bureau of Investigation and local law-enforcement laboratories, should be incorporated into the system. The diagnostic results in the databases should be screened continuously for possible newly emerging or unusual patterns of disease and infestation.

Early detection and diagnosis are pivotal for limiting the extent of an outbreak. Early detection and response at the local and state levels are particularly important. Technology must be rapid, field-deployable, accurate, and sensitive

---

[6]On May 30, 2002, USDA announced that, out of its allocations for homeland security under the fiscal year 2002 defense-spending bill (HR 3338), 1) $20.6 million will be provided to state and university cooperators to establish a network of diagnostic laboratories for animal diseases, and 2) $4.4 million will be used to improve plant pest and disease diagnostic capabilities. These actions appear to be consistent with the committee's recommendation.

and should be inexpensive and require little training for use. There have been substantial advances in technology for detection and diagnosis of microbial diseases, but they have often not included agriculturally important pests or pathogens, nor have they been inexpensive or field-deployable. Therefore, the committee recommends a large expansion in detection and rapid diagnostic capabilities for high-priority pests and pathogens.

In some cases, available detection and diagnostic tests are not used because they have not been validated and accepted by the international community. The committee recommends national—and international—mechanisms to promote the testing and validation of emerging detection methods and diagnostic tests. The mechanisms should be integrated into the laboratory-response network described above.

Exchanging information during crises is essential for coordinating responses, mobilizing resources, preventing inappropriate actions, and ultimately minimizing impacts. Federal, state, and local agencies, industry, producers, and academic scientists need to communicate effectively with each other and be informed on a real-time basis so that threat agents can be intercepted and eliminated, or at least controlled as soon as possible. Recent test exercises involving simulated release of bioterrorism agents highlight the lack of communication among agencies and partners as an important deficiency.

Public communication is vital to ensuring the effectiveness of mitigation measures, such as quarantines and destruction of crops, and minimizing exposure to the threat agents. There is no national communication, data-management, and information system for agricultural bioterrorism. The committee recommends that one be established and patterned after one being designed for human-health threats, the Center for Disease Control and Prevention's (CDC) Health Alert Network. That network is a nationwide, integrated information and communication system that serves as a platform for distribution of health alerts, dissemination of prevention guidelines and other information, distance learning, national disease surveillance, and electronic laboratory reporting, as well as for strengthening preparedness at the local and state levels. The committee also recommends making the agricultural system interactive with this existing CDC system.

The committee did not analyze the potential effectiveness of the network in comparison with other alternatives for improving communication, but recommends that the approach for agricultural bioterrorism take full advantage of the efforts already under way for human health. The recommended system should include a directory of experts in related fields and relevant data from R&D projects, field stations, regional agriculture laboratories, and agriculture professionals. The information system should be somehow integrated with the laboratory network suggested above.

Knowledge of the overall system for response to and recovery from an agricultural disease or infestation is vital for knowing which elements to test further and fortify. The committee recommends that pathway analyses for a set of potential threat agents be developed to illuminate weaknesses and correct them before an attack occurs. Each analysis should involve a description of the strengths and weaknesses of the system for the specific pest or pathogen and should propose points of intervention or risk mitigation. The committee recommends that model scenarios of mock attacks be exercised with the draft pathways so that important information about strengths and deficiencies in the defense system can be gleaned.

During a biological attack, any delays in implementation of eradication or management plans allow further spread of diseases or infestations, economic damage, or harm to public, plant, or animal health. Therefore, the committee recommends having plans for selected threat agents ready before an attack occurs in order to assist in smooth implementation of a response. These plans should be tested in advance for validation purposes.

Previous responses to agricultural diseases or infestations have highlighted communication problems among various professional sectors. For example, during the West Nile virus outbreaks in 1999, wildlife-health experts had difficulty connecting with the public-health community, and vice versa. In addition, USDA's current published emergency-response plans are not designed for intentional threats and therefore do not factor in the role of the law enforcement community. It is unclear in some cases whether eradication and management or investigation and law enforcement would take priority. Therefore, the committee recommends that response plans for various biological threats define clearly the roles and priorities of the agricultural, wildlife, public-health, intelligence, and law-enforcement communities.

As discussed above **(Finding I.C)**, international cooperation is important for the control, management and eradication of disease or infestation. A threat to the United States will always exist if a threat agent can be found in another country. The committee therefore recommends an effort, on the part of the US government and agricultural community, to assist in controlling agricultural threat agents that occur naturally in other countries.

Credible public information and effective communication are vital for effective management of bioterrorism and for reducing an attack's consequences. As a partner in any response effort, the public should be kept well-informed. Responsible agencies should provide a clear and consistent message about the nature and extent of the threat so as to facilitate mitigation measures and minimize economic, social, psychological and health effects. The committee recommends programs that incorporate the following three elements critical for effective communication: an identified lead spokesperson for each class of agents (ideally a person who is an expert on the agents, an effective communicator, and involved in federal response), a public and mass-media information plan for each agent or type of agent, and training of local officials in effective public communication.

The committee did not find a comprehensive educational program involving cooperative extension, academic institutions, professional or scientific societies, industry representatives, and commodity groups and aimed at increasing the awareness of pests and pathogens new or foreign to US agriculture. The committee recommends the development of such a program to help to improve early recognition of agricultural threats, especially at the state and local levels. This program should include diagnostic information and illustrative material, and it should specify whom to call in the event of a suspicious case or discovery. The program should be tailored to likely front-line personnel (such as, farmers, field agronomists, elevator operators, local veterinarians, and livestock buyers) as the primary audience, but should also include public and private diagnostic and research laboratories at the regional, national, and international levels to increase awareness to the greatest extent possible.

Recommendation I.F

As previously stated, the dynamic nature and extent of the threats to our agricultural economy will not allow us to plan for every specific agent or contingency **(Finding I.G)**. However, over the long term, a sound technical base will contribute to technology development and education for both general agricultural health and agricultural-bioterrorism preparedness and deterrence. Therefore, the committee recommends enhancing basic-research programs in the biology and epidemiology (including molecular epidemiology) of agricultural threat agents, the pathogenesis of crop and livestock diseases, and the social and behavioral aspects of agricultural bioterrorism. Information obtained from those programs should lead to better tools for preventing, detecting, responding to, and recovering from biological attacks.

Tools that are needed include novel inspection technologies for use at borders and more-rapid detection and identification technologies. In the short term, there are needs to identify animal threat-agent outbreaks that can be affected by vaccination and to analyze the cost-effectiveness of developing additional vaccine stockpiles for those agents. The committee recommends that exploration and development of newer vaccine technologies (such as DNA, vectored, and replicon systems) be pursued.

Genomic information has an important role in bioterrorism defense. DNA-sequence information has the potential to lead to determinations of the biogeographic lineages of pathogens and pests, which can help to determine the sources of pathogens. Such information can be critical when attribution is necessary for response. Research on information systems to interpret and use such information should be enhanced.

Finally, there is a pressing need to study social and psychological dimensions of agricultural terrorism. The goals include: 1) improving our ability to profile potential perpetrators and use more-effective intelligence methods, 2) developing more-effective education and crisis communication approaches, 3) better understanding public concerns, 4) increasing public understanding of and cooperation with needed mitigation measures, and, most important, 5) helping affected individuals, families, communities and regions cope with the consequences.

Through Chapter 4 and in Chapter 3, the committee identifies examples of natural and social-science R&D directions that are needed to enhance our ability to defend against agricultural bioterrorism. Resources are always limited, and not all avenues can be pursued. Therefore, R&D priority-setting is a critical part of an overall defense plan. The committee recommends that coordinated, national mechanisms for doing so be explored.

Recommendation I.G

Industry has a critical role to play in prevention of and response to a biological attack against agriculture. However, the priorities of the public and private agricultural sectors might differ in a biological attack. There are also needs for better incentives and information for private-sector development of technologies for countering agricultural bioterrorism that emerge from academic and government research. Much industrial R&D is proprietary and not available to the government via the literature or at open scientific meetings. Furthermore, industry sectors with applicable technologies might not know the government's needs and plans for countering agricultural bioterrorism.

The committee did not explore alternative mechanisms to facilitate public-private cooperation. However, the status quo is not optimal, because it tends to confine both government and industry in reactive rather than proactive modes. The proactive mode is likely to provide greater deterrence or more-rapid response. Therefore, the committee recommends that a public-private advisory council be established to act as a mechanism for building relationships, exchanging information, planning cooperative programs, and improving technology transfer between public and private sectors. The committee recommends this as an important, low-cost first step through which different or larger initiatives could be proposed.

CONCLUDING REMARKS

Agricultural bioterrorism anywhere will have, in most cases, worldwide consequences because of the global nature of the agricultural system. Consequences will include disruption of markets, difficulties in sustaining an adequate food and fiber supply, and the potential spread of disease and infestations throughout the nation and around the world. Significant adverse public health effects could occur through the use of contagious or particularly devastating zoonotic agents. Furthermore, not only must our food and fiber supply be safe and secure, but the public must have confidence in it.

The US system for defense against intentional biological threats to agricultural plants and animals uses existing infrastructure designed to protect against unintentional threats. Strengthening that existing system is recommended by the committee as a resource-efficient, as well as effective, part of the defense against terrorism. However, without the recommended comprehensive plan and additional capabilities designed specifically for defense against intentional threats, a strengthened system would not be sufficient. On the other hand, most of the improvements that are needed to adapt the system to the terrorist threat will also strengthen the system's ability to protect against unintentional threats. Thus, a system developed to counter agricultural bioterrorism will improve the security of the US food and fiber system, even if the terrorist threat recedes.

Some of the committee's recommendations require more development time and resources than others and therefore are intermediate to long term for full implementation. The committee suggests that the following recommendations can be achieved in the near term and should have priority for immediate action:

- Establish and train credible spokespersons for classes of threat agents.
- Define legal and jurisdictional authority and lead roles at the federal, state, local and private levels. Include specifications for interagency cooperation.
- Define a categorical priority list of threat agents for planning.
- Develop and exercise model scenarios of agricultural bioterrorism attacks.

Among those recommendations that require more time for implementation, the committee suggests that the following have priority and that plans for their implementation begin immediately:

- Increase basic understanding of pathogenesis in plants and animals.
- Establish a laboratory response network for detection, identification and diagnosis.
- Establish a nationwide system for communication, data management and analysis.

1

Introduction

BACKGROUND

In the wake of the events of September 11, 2001, few would disagree that the United States, more than ever before, must be alert to and prepared for the possibility of surreptitious attacks within its borders—attacks aimed not so much to achieve strategic military victories as to cause indiscriminate destruction, economic disruption, widespread injury, fear, uncertainty, and the social breakdown. Such attacks are generally defined as terrorist attacks, or terrorism; if the weapon used involves a biological species (biological threat agent), it is bioterrorism.

The terrorist attacks on the World Trade Center and the Pentagon in September 2001 made it clear that some groups have the organizational skills and the motivation to carry out catastrophic strikes. The later deliberate distribution of anthrax spores, although at this point not as clear in origin or intent, was for some a confirmation of—and for many Americans an awakening to—the danger of "nontraditional weapons", particularly biological weapons.

The events have also provided an illustration of the importance of a carefully structured and coordinated defense and response plan. They have made clear how necessary it is to be able to respond, to be perceived as able to respond, and to help the public to have the perspective to understand the threat—neither to underestimate its danger nor to overestimate its consequences. Indeed, it is often the case where terrorism, and bioterrorism in particular, is concerned that the public's perception of risk or danger is at least as great a concern as, and in some cases greater than, the tangible or material danger itself. Where fear and uncertainty are the aims, a credible hoax can also achieve a terrorist's goal. Thus, with respect to bioterrorism, effectively communicating with the public and maintaining its trust is crucial. Timeliness, achieving the right balance between information protection and information sharing, and navigating the complex world of electronic and print media are all vital components of proper communication.

The study reported on here was undertaken to evaluate the nation's preparedness to deal with bioterrorist threats to agriculture—that is, the ability of the United States to deter or prevent an intentional biological attack against its supply of food and fiber or, if that is impossible, to detect, thwart, respond to, and recover from an attack (Box 1-1). The study was initiated several months before September 11th, 2001, on the basis of a recommendation made by a National Research Council (NRC) steering committee that met in Washington, DC, in August 1999 (Appendix A). The steering committee noted that although the growing threat of biological weapons was widely recognized, most of the focus had been on the threat of species that directly affect humans, such as anthrax and smallpox. There had been little discussion of possible attacks on plants and animals.

Historical evidence suggests that a number of nations have considered using or have even used biological agents against plants and animals. During World War I in the United States, German agents infected horses with bacteria that cause glanders, a fatal equine and human disease. The Soviet Union military also used glanders in the early 1980s during war in Afghanistan (Wilson et al. 2000). In World War II, the United States, Great Britain, and others all had offensive biological-warfare programs directed against plants or animals, and these continued after the war until 3 years before the adoption of the Biological and Toxic Weapons Convention of 1972 (BWC). Some countries continued their offensive programs after 1972. Iraqi bioweapon development in the late 1980s and early 1990s included anticrop efforts based on wheat smut, a disease caused by a fungus in the genus *Tilletia*. There is also evidence that the Soviet Union had an extensive agricultural-terrorism program aimed at animals and plants (Alibek 1999a). The program, code-named Ecology, developed variants of agents to attack cattle, swine, and poultry. The agents were contagious, so that even if only a few animals were infected initially, the diseases would eventually spread, causing severe consequences, and they were designed to be sprayed from low-flying airplanes (Alibek 1999b). Recently, the US has been involved in research and development for the use of the fungus, *Fusarium*, for destruction of coca plants in Colombia. Although many believe that this is not a direct violation of the BWC, there is controversy over this approach, due in part to concern about inducing epidemics in other plants and transferring knowledge about such approaches to terrorists (Rogers et al. 1999).

Toxins and chemicals have also been used in terrorist attacks against livestock. In 1952, the Mau Mau, a nationalist liberation movement in Kenya, poisoned 33 steers at a British mission station, using a local toxic plant known as African milk bush, (*Synadenium compactum* 'Ruby') (Wilson et al. 2000, Center for Non-proliferation Studies 2001). Attacks using chemical agents have recently been directed at US agriculture. In December 1996 in Wisconsin, an unidentified

BOX 1-1
Terms Used to Describe Strategies for Avoiding or Responding to a Biological Attack

A number of strategies for avoiding or responding to a biological attack come into play separately or in a coordinated way at various stages. These are the terms we use in this report to describe the strategies:

Deterrence. Efforts before any attempted biological attack to decrease the likelihood that an individual or group will mount such an attack. Deterrence can include plans and actions that discourage a potential attacker about the possibility of success or cause fear of capture or retribution.

Prevention. Defense plans or defenses that stop an individual or group from carrying out an attack. Defense can include disruption of financial and communication infrastructures, disruption of or attempts to control the availability of disease- or infestation-causing biological agents, or efforts to reduce support of potential attackers by addressing causes of social and political unrest. It can also involve intelligence, detection of an impending attack, and preemptive action to avoid an attack.

Detection. Technologies and protocols to locate and identify a biological agent by some combination of direct characterization and recognition of disease or infestation symptoms while the agent is in transit or during or after its release.

Thwarting. Action taken once a biological attack or impending biological attack is detected to prevent the actual infection or infestation.

Response. Plans and actions to combat, control, and limit the consequences of a biological attack plus actions to identify and appropriately deal with the perpetrators.

Recovery. Cleanup, restocking, restarting of production, and helping affected individuals, families, communities, and regions to return to normal.

perpetrator used chlordane, an organochlorine pesticide, to contaminate animal feed (Neher 1999). Chlordane accumulates in the fat of animals and can adversely affect consumer health at very low concentrations. The response was well managed by federal, state, and local agencies with good cooperation by the feed industry. Within 2 days of detection, potentially contaminated animal products were recalled and the feed replaced. However, the attack caused considerable fear and panic and led to the disposal of feed and fat with an estimated value of nearly $4 million.

Attacks have also been directed at research facilities and research plots where work with laboratory animals or genetically modified plants is conducted. The fire-bombing of university laboratories and uprooting of plants in research plots, which often destroys years of research aimed at peaceful uses of plants and animals, can be just as destructive as terrorist acts that originate outside our borders.

These real-world examples of terrorism aimed at plants and animals provide a baseline of experience and point to vulnerabilities. In light of this history, the steering committee recommended that an NRC study be undertaken to investigate the threat of biological attacks against America's agricultural sector and the nation's preparedness to deal with them. The US Department of Agriculture Agricultural Research Service (USDA-ARS) accepted the recommendation and provided the financial support for this study of biological threats to agricultural plants and animals (Box 1-2).

There are obvious similarities between the biological threats to people and the biological threats to plants and animals. Both involve the surreptitious delivery of a biological agent, which is difficult to prevent because of the ease with which agents can be transported. In both cases, a major challenge to the perpetrators is to find ways of effectively dispersing the agent, and major challenges to those reacting to the attack (if deterrence or prevention fails) are early detection and effective containment (which are related). Identification of the perpetrators and detection of the attack can be hampered in both situations by a period of latency, when there are no easily observable symptoms of infection. The period of latency can last from several days to a year or more. Adding greatly to the lag in detection is the time needed for discovery of plants or animals with symptoms. Because the surveillance of crops in particular is minimal, it can take several years before an infected plant is discovered after pathogen introduction.

Some of the social and psychological dimensions of agricultural bioterrorism are likely to be similar to those of terrorist attacks directly on humans, particularly if the threat agent used is a human, as well as a plant or animal, pathogen (for example, fear for human life and health, fear generated by demonstrated terrorist ability to penetrate defenses, confusion arising from conflicting expert opinion, and loss of confidence in public officials). However, some effects, such as those which are economic, will differ in nature and distribution (for example, concern for animal welfare, widespread disruptions of personal and business

BOX 1-2
Statement of Task for Study

In order to evaluate US preparedness for biological threats directed towards agricultural plants and animals, the United States Department of Agriculture asked the National Research Council's (NRC) Board on Agriculture and Natural Resources to convene a diverse group of experts to evaluate the ability of the United States to deter, prevent, detect, thwart, respond to and recover from an intentional, biological attack against the nation through its supply of food and fiber. This NRC study will emphasize deterrence and prevention and will include an analysis of the defensive posture before an attack takes place as well as capability to respond to an attack. The study will focus on two key areas and how these two areas are intertwined:

1) **US System.** A selected set of biological agents, representative of various types of potential threats, will be used to examine, under different scenarios, the probable response and efficacy of the current agricultural protection system and related health, law enforcement and intelligence systems. The study will identify the key organizations and participants in these systems at federal, state, and local levels, and will evaluate their respective roles and critical interactions to defend against an attack in the first place and to respond in the event of a biological attack. These organizations will include public health agencies, the intelligence community, law enforcement and other agencies that will need to partner with the US agricultural system. Short-term (e.g., government responsiveness) and long-term (e.g., research and education) national defense strategies will be evaluated.

2) **Use of Science in the US System.** The study committee will review the use of scientific information in providing the basis of defense policy and procedures to defend US agriculture, and will explore the potential of microbial forensics, vaccine/drug development, biochemical detection, decontamination technology, molecular epidemiology, and other technologies as tools to deter, prevent, thwart or resist a biological threat. Scientific research needed to enhance agricultural defense capabilities will be identified where appropriate.

The study committee will make recommendations, if necessary, on ways to improve the defensive posture, national security and scientific capabilities of the US agricultural and food system and the preparedness of federal agencies to catalyze action, cooperate, and effectively respond in order to protect the nation's supply of food and fiber.

travel resulting from quarantines in the countryside, terrorist motivation in target and threat agent selection, and lack of public understanding and acceptance of extraordinary practices required for protection and recovery of agriculture). Also, in the case of agricultural bioterrorism, psychological effects include a loss of confidence in the safety of the food supply. The challenge to the government, working with food producers and distributors, consumer groups and the media would be to appreciate public concerns and help the public understand the steps being taken to ensure food safety. Openness about the magnitude of risk, if any, would be required.

There could be geographic differences between the two types of biological attacks. Biological attacks on people seem likely to occur primarily in urban areas, but attacks on agricultural stocks seem likely to occur in rural areas. The infrastructures required to respond to (or prevent) an attack can be expected to be quite different in the two kinds of settings, given the different population densities, communication systems, and even physical landscapes. Urban and rural settings also present different social and personal circumstances.

How those differences would play out under the stress of an attack is not easy to assess in advance. For example, the social coherence of rural communities and the attachment to place may be useful in organizing communication networks and a community response to an attack. But, fear and uncertainty with respect to one's livelihood, property, and even identity in the rural setting, in contrast with fear and uncertainty with respect to one's personal safety in the urban setting, may generate different sets of emotions and different levels of cooperation. Indeed, the more complex responses to agricultural epidemics have been apparent to those who have had to implement quarantines and destroy stock in response to natural outbreaks. For example, some growers' reluctance to report an exotic pest or pathogen, especially if they know that the pest or pathogen is federally quarantined, could adversely affect their ability to sell product, or lead to the destruction of their herd or crop.

In many ways, attacks on plants and animals may be easy to mount. Agricultural crops and animals are often grown, housed, or grazed in relatively high-density and uniform conditions, which make the spread of disease and infestations more rapid and effective. Exotic pests and pathogens, often hosted and sustained by native wildlife and plants, may affect agricultural stock. In addition, wildlife and native plants might be targets of terrorism, given their aesthetic importance to and use by people. Genetic homogeneity, often desirable in agriculture to optimize yields or nutritional content, adds to the vulnerability of crops and animals to epidemics. Vaccination is difficult and problematic even in humans, and in the case of most agricultural species, has the additional problem of usually being economically impractical.

Furthermore, many diseases and infestations can cause, and in recent years have caused, widespread plant and animal epidemics. The most recent example is the outbreak of foot and mouth disease (FMD) in the United Kingdom in 2001.

Thus far, that epidemic, with about 2,000 confirmed cases of infection, has required the destruction of close to 4 million animals and has resulted in direct and indirect economic losses variously estimated at $6-$30 billion (Ferguson et al. 2001 and Becker 2001b). Other important epidemics have occurred during the last 2 decades. Among them are the following:

- Avian influenza. A serious avian influenza epidemic occurred among poultry in the northeastern United States in 1983-1984. About 17 million birds had to be destroyed over a 2-year period at a cost of nearly $65 million (USDA-APHIS 2001c).
- Nipah virus. This zoonotic virus, carried by pigs, first appeared in Malaysia in 1998. It infected 257 people, 100 of them fatally. Containing it required the destruction of 900,000 pigs, more than one-fourth of Malaysia's stocks (USDA-APHIS 1999a).
- Hog cholera. This highly contagious viral disease appeared in the Netherlands and Belgium in 1997 and reappeared in Haiti and the Dominican Republic, which had eradicated it in the early 1980s (USDA-APHIS 1998a). Those appearances led to fears that it could reinfect American animals.
- Karnal bunt. This fungus first appeared in the southwestern US wheat crops in 1996 and necessitated quarantines in an industry in which exports amount to $3.5 billion per year (USDA-APHIS 2001d). There is evidence of substantial regional economic damage due to this fungus (Bevers et al. 2001). For example, the 2001 outbreak of the disease in 4 counties of north Texas caused $15 million in losses in the outbreak area.
- Citrus canker. An outbreak of Asiatic citrus canker, which apparently started near the Miami airport in the middle 1990s, had spread by 2001 to an area of more than 500 mi^2 and has required the removal of 1.5 million citrus trees at a cost of over $200 million (Schubert et al. 2001).
- West Nile virus. This zoonotic virus first appeared in the Western Hemisphere in 1999, and, in addition to infecting a number of horses, had by the end of 2000 infected about 83 people, nine fatally (CDC 2001a). It is not known how far it will spread.
- Mediterranean fruit fly (Medfly). Infestations between 1980 and 2000 in California and Florida were eradicated at an estimated cost of $400 million. They engendered major problems for USDA-APHIS and state plant health officials because of opposition to the use of malathion over residential areas by airplanes and helicopter. The largest Medfly infestation in recent history was in 1980-82 in the San Francisco Bay area of California, which involved aerial malathion bait sprays over 1,400 square miles, and was eradicated at a cost of about $100 million (CDFA 2002).

The methods used to respond to and recover from a biological attack depend on whether plants and animals or humans are attacked. Large-scale destruction of crops or herds to control the spread of the disease or infestation is one option available in agriculture. Others are complete quarantine and removal of agricultural products from the market for long periods—approaches impractical for human populations if they are applicable at all.

Those response and recovery approaches themselves can have serious economic consequences. The entire food and fiber system—including farm inputs, processing, manufacturing, exporting, and related services—is one of the largest sectors of the US economy, accounting for output of $1.5 trillion, nearly 16% of the gross domestic product, and 17% of the civilian labor force (USDA 2001a, USDA-ERS 2002a). Although activity is not uniform throughout the country, it is widespread. California leads the nation in the value of final agricultural output, producing over 13% of the nation's output in 2000, followed by Texas, Iowa, Nebraska, Kansas, Minnesota, North Carolina, Illinois, Florida, and Wisconsin (USDA-ERS 2001a) (Table 1-1).

The agricultural economy accounts for 8% of US exports and 3% of imports (USDA-ERS 2002a), and trade in agricultural commodities has increased markedly during the last decade (Tables 1-2 through 1-5). The largest share of imports is horticultural products, including fruits, nuts, and vegetables (Table 1-5); and the imports come from countries spread around the globe (Table 1-3). This substantial international trade is both a protection and a vulnerability for the United States. On the positive side, it makes it very unlikely that a biological attack on US agriculture would lead to serious food shortages. Some products might become unavailable, and food costs would likely increase, but the well-developed international market means that the infrastructure for importing and distributing food should make it possible to continue to meet the basic needs of the American public.

On the negative side, although the international trade in agricultural commodities would likely help the United States to avoid food shortages in the event of a biological attack, lost trade in international markets contributes to economic costs resulting from the attack, particularly the indirect costs. In 2001, the total value of US agricultural exports was nearly $50 billion, including substantial shares from a wide array of commodities (Table 1-5). The direct costs associated with an incident would be much the same whether or not exports were a large fraction of sales. They would include the economic value of losses in product and capital stock due to the incident. Other costs would arise in controlling or containing the results of the attack, and in cleaning up after its effects. If there were a need for broad scale environmental cleanup, these latter costs could be high. In addition, transfer costs would be associated with reimbursing and compensating producers.

The indirect costs are likely to be even larger, but they are harder to predict because they depend heavily on market effects, and here the international response could be critical. Clearly, there would be the immediate costs of lost market (both international and domestic). Indeed, some producers of substitute products

TABLE 1-1 Leading States in Cash Receipts for Top 25 Commodities, 2000

			Top 10 States and Value of Cash Receipts, Millions of Dollars									
Commodity[a]	Rank	Value, millions of dollars	1	2	3	4	5	6	7	8	9	10
All commodities	—	193,586	CA 25,510	TX 13,344	IA 10,774	NE 8,952	KS 7,905	MN 7,522	NC 7,410	IL 7,022	FL 6,951	WI 5,221
Livestock and products	—	99,473	TX 9,162	CA 6,269	NE 5,923	IA 5,747	KS 5,488	NC 4,275	MN 3,875	WI 3,804	OK 3,441	CO 3,332
All crops	—	94,113	CA 19,241	FL 5,573	IL 5,312	IA 5,027	TX 4,181	MN 3,647	WA 3,339	NC 3,135	NE 3,029	IN 2,886
Cattle and calves	1	40,761	TX 6,815	NE 4,948	KS 4,948	CO 2,551	OK 2,298	IA 1,840	SD 1,411	CA 1,267	MO 1,045	MT 966
Dairy products	2	20,622	CA 3,704	WI 2,690	NY 1,544	PA 1,520	MN 1,127	TX 766	ID 762	MI 729	WA 711	NM 644
Corn	3	15,086	IA 2,656	IL 2,582	NE 1,725	IN 1,298	MN 1,173	OH 733	KS 692	SD 543	WI 542	MO 524
Broilers	4	13,953	GA 2,029	AR 1,927	AL 1,748	NC 1,418	MS 1,221	TX 880	DE 497	CA 469	MD 462	VA 441

continues

TABLE 1-1 Continued

Commodity[a]	Rank	Value, millions of dollars	Top 10 States and Value of Cash Receipts, Millions of Dollars									
			1	2	3	4	5	6	7	8	9	10
Greenhouse and nursery[b]	5	13,037	CA 2,778	FL 1,548	TX 1,179	NC 987	OR 654	OH 554	MI 491	PA 312	NJ 297	NY 295
Soybeans	6	12,540	IA 2,166	IL 2,140	MN 1,320	IN 1,140	OH 832	NE 800	MO 755	SD 666	AR 412	MI 354
Hogs	7	11,772	IA 3,071	NC 1,648	MN 1,207	IL 826	NE 683	IN 592	MO 591	OK 473	OH 330	KS 295
Wheat	8	5,470	KS 891	ND 740	MT 410	WA 395	OK 326	SD 299	ID 266	MN 249	CO 202	TX 172
Cotton	9	4,555	TX 1,027	CA 807	GA 411	MS 406	AR 392	NC 347	LA 249	TN 196	AZ 187	MO 148
Chicken eggs	10	4,347	GA 370	OH 342	AR 314	PA 287	IN 262	AL 260	TX 257	IA 241	CA 238	NC 222
Hay	11	3,408	CA 441	ID 236	WA 184	TX 168	NM 162	OR 158	CO 155	PA 137	KS 133	SD 115
Grapes	12	3,104	CA 2,836	WA 127	NY 46	OR 26	MI 24	PA 17	AZ 14	GA 4	NC 3	OH 3

Turkeys	13	2,786	NC 434	MN 361	MO 272	VA 238	AR 219	CA 211	IN 157	SC 141	TX 115	PA 114
Potatoes	14	2,469	ID 598	WA 408	WI 176	CA 176	OR 136	MN 115	ME 115	ND 112	MI 106	CO 100
Tobacco	15	2,315	NC 854	KY 674	TN 200	SC 168	GA 150	VA 132	OH 28	FL 25	IN 21	MD 15
Oranges	16	2,052	FL 1,397	CA 645	AZ 7	TX 4	na	na	na	na	na	Na
Lettuce	17	1,862	CA 1,484	AZ 349	NJ nr	CO 11	FL 4	na	na	na	na	na
Tomatoes	18	1,823	CA 951	FL 507	TN 35	OH 35	GA 33	IN 33	NJ 32	VA 31	NY 31	PA 28
Apples	19	1,453	WA 822	NY 127	CA 102	MI 89	PA 55	VA 40	OH 22	NC 18	WI 17	OR 16
Sugar beets	20	1,215	MN 339	ID 241	ND 195	CA 112	MI 112	MT 53	WY 45	CO 38	NE 37	WA 26
Horses and mules	21	1,156	KY 1,040	NJ 116	na	na	na	na	na	na	na	na
Rice	22	1,151	AR 549	CA 232	LA 149	TX 83	MS 81	MO 57	na	na	na	na

continues

TABLE 1-1 Continued

			Top 10 States and Value of Cash Receipts, Millions of Dollars									
Commodity[a]	Rank	Value, millions of dollars	1	2	3	4	5	6	7	8	9	10
			CA	FL	OR	NC	PA	NY	WA	MI	VA	OH
Strawberries	23	1,014	767	168	17	17	7	7	7	7	6	5
			FL	LA	HI	TX						
Cane for sugar	24	914	442	345	82	46						
			MS	AR	AL	FL	ME	WA	LA	ID	VA	CT
Aquaculture	25	844	302	96	82	77	62	45	44	38	25	18

[a]Twenty-five leading commodities ranked by value of farm marketings.
[b]Excludes mushrooms.

SOURCE: US Department of Agriculture, Economic Research Service, 2001a.

TABLE 1-2 US Agricultural Trade and Exports by Major Commodity Group

Item	January-December 2000	January-December 2001	Change '01>'00
	Million dollars		Percent
Agricultural exports	51,296	53,795	5
Imports (c.i.f.)	42,300	42,566	1
Imports (customs value)	38,982	39,377	1
Trade balance			
Exports minus c.i.f. imports	8,996	11,228	25
Exports minus customs-valued imports	12,314	14,417	17
Exports by major commodity group			
Live animals	825	854	4
Red meats and products	5,276	5,294	0
Poultry meats and products	2,134	2,534	19
Dairy products	1,018	1,130	11
Hides and skins	1,562	1,980	27
Animal fats and other products	835	718	–14
Wheat	3,354	3,355	0
Rice	855	716	–16
Corn	4,469	4,497	1
Other feed grains	700	731	5
Other grain products	1,753	1,880	7
Animal feeds and oil meal	3,782	4,219	12
Soybeans	5,258	5,420	3
Other oilseeds	844	1,115	32
Vegetable oils	1,259	1,255	0
Fruits, nuts, and products	4,061	4,066	0
Vegetables and products	4,457	4,485	1
Juice, wine, and beverages	2,004	2,051	2
Cotton and linters	1,893	2,176	15
Tobacco, un-manufactured	1,204	1,268	5
Sugar and tropical products	1,649	1,860	13
Other	2,105	2,190	4
Total Exports	51,296	53,795	5

SOURCE: US Department of Agriculture, Economic Research Service, 2002d.

may benefit from increased demand. The larger question involves long-term market behavior. The major swings in product availability (and price) in the short term introduce the potential for destabilization of future markets. The question is how long it would take and, indeed, whether it would even be possible to regain public confidence in the affected commodity. The international market presents a great challenge in that respect, and the restrictions placed on the markets, even if lifted later, might have long-term effects—including loss of export markets or elimination of commercial production. The 1997 outbreak of

TABLE 1-3 Value of US Agricultural Imports, by Country of Origin

	Value, Billions of Dollars	
Country of origin	Average: 1990–1996	Average: 1997–2000
Canada	4.6	7.9
European Union	5.3	7.5
Mexico	2.9	4.6
Brazil	1.3	1.4
Indonesia	1.0	1.3
Colombia	0.9	1.3
Australia	1.0	1.2
New Zealand	0.8	0.9
Chile	0.5	0.9
Costa Rica	0.5	0.8

SOURCE: US Department of Agriculture, Economic Research Service, 2001b.

TABLE 1-4 Value of US Agricultural Exports, by Importing Country

	Value, Billions of Dollars	
Importing country	Average: 1990–1996	Average: 1997–2000
Japan	9.3	9.6
European Union	7.6	7.7
Canada	5.1	7.0
México	3.7	5.8
South Korea	2.6	2.7
China (Taiwan)	2.2	2.2
China (Mainland)	1.1	1.4
Hong Kong	1.0	1.4
Egypt	0.9	1.0
Russia	1.0	0.9

SOURCE: US Department of Agriculture, Economic Research Service, 2001b.

foot and mouth disease in Taiwan, for example, curtailed Taiwanese exports to Japan and had a serious effect on pork production in Taiwan. Before the outbreak, Taiwanese pork exports represented 40% of the frozen-pork market and 68% of the fresh-chilled-pork market in Japan. After the outbreak, Taiwan's share of Japan's import market went to other sources (including the United States) (Fuller et al 1997, USDA-APHIS 1998b). The tendency of people, communities and the nation as a whole to draw together in the face of a terrorist attack would help the process of rebuilding public confidence. However, this factor would likely not come into play in the international market.

TABLE 1-5 Value of US Agricultural Imports, by Commodity Groups

Item	January-December 2000	January-December 2001	Change '01>'00
	Million dollars		Percent
Imports customs value by major commodity group			
Live animals	1,865	2,170	16
Red meats and products	3,774	4,175	11
Poultry meats and products	263	248	–6
Dairy products	1,671	1,789	7
Other animal products	764	758	–1
Grains, feeds, and oil meal	3,237	3,464	7
Fruits, juices, and nuts	5,373	5,257	–2
Vegetables and preparations	4,740	5,252	11
Oilseeds	297	278	–6
Vegetable oils	1,399	1,219	–13
Wine	2,207	2,250	2
Malt beverages	2,179	2,348	8
Coffee	2,700	1,677	–38
Cocoa	1,404	1,535	9
Rubber, natural	842	613	–27
Sugar, cane, and beet	512	516	1
Sugar products	1,043	1,086	4
Tobacco, un-manufactured	628	680	8
Other tropical products	1,685	1,619	–4
Other competitive products	2,397	2,443	2
Total Imports	38,982	39,377	1

SOURCE: US Department of Agriculture, Economic Research Service, 2002b.

Another distinction between biological attacks on humans and biological attacks on plants and animals is that it may be more difficult in the case of agricultural commodities to determine whether the problem is natural or accidental, or the result of a deliberate biological attack. For example, smallpox has been all but eliminated from the world except for a few stockpiles. A natural outbreak would be virtually impossible, and any appearance of the disease would lead immediately to the conclusion that the virus had been spread by biological terrorists. Anthrax is more prevalent, but not in the particle size in which it can spread easily and infect the lungs, and not in the mail system or in densely populated areas. Therefore, when it is found as an inhalation disease, a biological attack is suspected.

Agricultural stocks, in contrast, are vulnerable to naturally occurring epidemic diseases and infestations and have been exposed to various kinds of outbreaks throughout the world in the last few decades. Moreover, plant pathogens and pests have seldom been eradicated, so there are often diseased or infested plant specimens in fields. That would appear to make it considerably easier for a bioterrorist group to obtain a suitable infectious agent and initiate an outbreak without raising immediate suspicion that the attack was deliberate. Some argue that the very fact that we have dealt with these diseases and infestations recently diminishes the seriousness of the problem; that is, for some agents, there is a considerable amount of knowledge about colonization and spread. We also know how to contain or mitigate the effects of some potential plant and animal diseases and infestations, although it is questionable how adequately this knowledge has been diffused to the "front lines" of pest or pathogen detection and control.

For various agricultural agents, there is significant variability in the time between introduction and manifestation of disease or infestation. Some agents, such as FMD virus, have short incubation times and rapid spread; whereas others, such as bovine spongiform encephalopathy (BSE), have very long incubation times. Perpetrators may choose different agents to achieve different goals. For example, if an agent with a long incubation period was chosen there would likely be greater time for perpetrators to escape from the geographic area of introduction before an attack was suspected. Difficulty in determining quickly whether an attack is deliberate introduces problems. If the attack is deliberate and is carried out by a knowledgeable group, the genetic form of the species used or the method of attack might lead to a sequence of events or a set of challenges somewhat different from that expected in an unintentional outbreak. For example, a deliberate multipoint release or the simultaneous release of several biological agents might require a different strategic deployment of field and laboratory analytical and human resources and even a different approach to containment from those required by a localized, single-species event. The implementation of a response plan based on the set of assumptions appropriate for an unintentional attack may be inadequate or even counterproductive. If the attack is deliberate, one important goal will be to identify and apprehend the perpetrators, and the response must deal simultaneously with the need to contain the outbreak and the need to facilitate the investigation of its causes. Proper containment will depend on the epidemiology of the disease or the ecology of the pest or pathogen and on the time that has elapsed since the introduction. The need for simultaneous efforts in containment and in investigation will affect not only the substantive procedures followed, but the selection of government organizations involved and the command and control structure to be invoked.

THE COMMITTEE'S ACTIVITY

In light of the foregoing, a set of questions needs to be answered to assess the nature of an intentional biological threat and the nation's preparedness for dealing with such a threat:

- How are intentional biological threats to plants and animals likely to differ from unintentional (natural or accidental) threats?
- How are intentional threats likely to increase substantially the probability of an epidemic outbreak; and are they likely to represent no more than a small increase in the existing probability of an unintentional threat?
- How or for what potential threats would a response plan for an intentional attack differ from a response plan for an unintentional event?
- Are existing scientific techniques adequate for the prevention, detection, and containment of and recovery from unintentional events adequate for intentional attacks? Are some scientific gaps ripe for investigation?
- Would addressing gaps in our technical capacity to prevent, detect, contain, and recover from unintentional events prepare us adequately to deal with intentional events?
- To what extent are the organizational structures and linkages—including administration, coordination, human resources, and technical tools—appropriate for deterring, preventing, detecting, responding to, and recovering from an intentional threat? Would addressing gaps in the organizational structure for responding to an unintentional event prepare us adequately to respond to an intentional event?

These questions provided a starting point for this committee in addressing its charge. The committee held a series of five meetings from May 2001 to January 2002. During public sessions at three of the meetings, formal presentations were made by representatives of federal departments and agencies and other experts (see Appendix B). Many documents—including official government publications, commentaries, and scholarly articles—were provided to the committee to inform its deliberations.

In organizing its report, the committee viewed it as useful first to review the existing US system for defending against biological threats to agricultural plants and animals. The results of this review are presented in Chapter 2. It then developed a set of scenarios to encompass different disease or infestation vectors, target plants or animals, ease of detection, time courses of outbreaks, and strategies for containment. The committee made a judgment about which potential threat agents would illustrate the broadest array of possibilities and important scientific and science-policy issues and, of those, which were well enough understood to permit useful examination. Lessons learned from the committee's in-depth examination are presented in Chapter 3.

On the basis of its analysis of the US agricultural defense system and the case studies, the committee tried to address the questions posed above and to identify ways in which the agricultural defense system could be improved (see Chapter 5). In Chapter 4 and throughout this report, the committee points out subjects of research that are ripe for investigation and could contribute to the strength and effectiveness of the defense system. The observations and recommendations in this report are addressed both to the government and to other organizations that have key roles to play in ensuring the nation's preparedness to deal with deliberate biological attacks on agricultural plants and animals.

2

The Current US System: An Overview

INTRODUCTION

The scope of the outbreak of foot-and-mouth disease (FMD) in Great Britain in 2001, the events of September 11, 2001, and the anthrax threats that followed have greatly heightened awareness in the agricultural sector of potential vulnerabilities of the US plant- and animal- health safeguarding system. The system in place to defend against threats to agriculture has weaknesses and needs (National Plant Board 1999, NRC 2002, NASDARF 2001, NISC 2001, NRC 1998), and the state of readiness to respond to intentional threats of agricultural terrorism is currently changing as federal, state, and local agencies work to improve their capacity. The secretary and deputy secretary of agriculture have undertaken a number of actions to strengthen the nation's ability to defend against biological threats to the nation's food and fiber supply (Veneman 2001a,b,c and Moseley 2001), and the US Department of Agriculture (USDA) has implemented a new structure for dealing with homeland security (Appendix D). The US Patriot's Act, signed on October 25, 2001, gives federal authorities wider latitude in the hunt for terrorists including broader surveillance powers, greater sharing of intelligence, data, and more authority to protect US borders.

Several additional bills have been introduced in Congress to combat terrorism, bioterrorism, and terrorism directed against US agriculture. They include provisions for upgrading animal and plant disease facilities in Plum Island, NY; Ames, IA; Athens, GA; and Laramie, WY; collaborative research with the Department of Justice; expanding USDA Animal and Plant Health and Inspection Service (APHIS), USDA Food Safety and Inspection Service (FSIS), and Food and Drug Administration (FDA) activities; biosecurity grants to academe and industry; planning for crisis communication and an education strategy; strengthening the food inspection system; state participation in foodborne-disease surveillance networks; studies of plant and animal disease; and better linkages to the intelligence communities. While this report was being prepared, the final version of the Public Health Security and Bioterrorism Preparedness and Response Act (HR 3448) passed the House and Senate. This bill authorizes $545 million for FDA and USDA to hire hundreds of new inspectors at borders, develop new methods to detect contaminated foods, work with state food safety regulators, and bolster safety at animal research labs (Tauzin 2002). The bill also gives more authority to FDA to detain suspicious foods and require prior notice of all food imports.

The fiscal year 2002 defense-spending bill (HR 3338) approved by Congress on December 20, 2001, includes over $300 million for USDA to support agricultural terrorism research and development, upgrade facilities, and improve border control (Kilman 2001, Malakoff 2002a). The bill includes $105 million for APHIS pest and pathogen exclusion, detection, and monitoring; $80 million for upgrading USDA facilities and operational security; $50 million for an animal biocontainment facility at the National Animal Disease Laboratory; $40 million for the USDA Agricultural Research Service (ARS); $23 million for the Plum Island Animal Disease Center; $15 million for security upgrades and bioterrorism protection for FSIS; and $14 million for increased security measures at the National Veterinary Services Laboratories in Ames, IA (USDA 2002a). Also, President Bush's proposed fiscal 2003 budget (as of February 1, 2002) includes a $146 million increase in funding to protect agriculture and the food supply: $48 million for animal health monitoring, including coordination and implementation of rapid response during outbreaks; $19 million for the Agricultural Quarantine Inspection program and improvements in technology and personnel at points of entry; $12 million for programs to expand technical services within APHIS for diagnoses, response, and management; $28 million for USDA-FSIS inspection and risk-management activities; $34 million for research aimed at protecting the food supply from attack; and $5 million to increase the capability of APHIS to monitor outbreaks of disease in foreign countries (USDA 2002a). While the above funding measures are consistent with strengthening the system for protecting US agriculture, it is beyond the committee's scope to perform cost-benefit analyses to determine their effectiveness or adequacy (See Executive Summary, and Chapters 1 and 5).

As enacted by Congress, the prevention, detection, eradication, control, suppression or retardation of the spread of animal diseases and plant pests new to or not widely distributed in the United States rests with the secretary of agriculture (US Congress 1884, 1903, 1962, 2000). Animal disease means any infectious or non-infectious disease or condition affecting the health of livestock or any condition detrimental to production or marketing of livestock (US Congress 2001). Plant pest means any living stage of any nonhuman organism that can directly or indirectly injure, cause damage to, or cause disease in any plant or plant product (US Congress 2000).[1] The Secretary has delegated the responsibility to APHIS.

The control of animal and plant pests and pathogens relies heavily on collaboration with international organizations such as the Food and Agriculture Organization and the Office of International Epizootics, other USDA units, other federal agencies, state departments of agriculture, other state agencies, industry groups, academe, professional societies, environmental organizations, community groups, and private individuals. It is clear that the federal government plays a key role in responding to and recovering from introductions of animal and plant pests and pathogens. Table 2-1 outlines current federal roles. The basis of the table was created by USDA (USDA-APHIS-VS 2001), and it was originally designed for unintentional introductions. It is important to note that the committee could not identify a similar one for intentional introductions of plant and animal pests and pathogens (Chapter 5, Finding I.B.1 and Recommendation I.C).

Efforts are under way to integrate federal animal-disease response planning more fully with the resources of state and local governments and institutions, the animal industry, volunteer organizations, and state emergency management organizations (USAHA 2002). For example, North Carolina has incorporated its interim Foreign Animal Disease Operations Plan into the Carolina Emergency Operations Plan. The operations plan outlines the actions and procedures that the State Emergency Response Team (SERT), in conjunction with USDA, will follow when a potential foreign animal disease is reported. The SERT is a team of state agencies, volunteer organizations, and state businesses that provide emergency and disaster support services to North Carolina (NCDA and CS 2002). The committee was unable to identify comparable initiatives for dealing with plant pests and pathogens at the state and local levels.

Risk assessment provides the scientific basis of regulatory and other intervention decisions. Predictive models and pest or pathogen, commodity, and pathway[2] analyses are applied to evaluate and predict the likelihood of occurrence of infestation and disease (Campbell and Madden 1990). The level of confidence in the assessments depends heavily on the availability and reliability of pathway and biological data, as well as the soundness of assumptions.

Authority for a federal response to animal diseases and plant pests "new to or not widely distributed in the United States" resides with APHIS, as delegated by the secretary of agriculture, but the initial detection of a new pest or pathogen on a farm or ranch, preliminary (and often scientifically authoritative) diagnosis, and development of a program for its control relies heavily—in many cases entirely—

[1]These are the statutory definitions of *plant pest* and *animal disease*. In this report, however, the committee uses *pest* primarily to signify a nonmicrobial agent (such as an insect) that causes plant or animal infestations, and *pathogen* to signify a microbial agent (such as a bacterium, virus, or protozoan) that causes plant or animal disease.

[2]Any means that allows the entry or spread of a pest or pathogen (FAO 1996).

TABLE 2-1 Current Federal Capabilities to Respond to Animal (and Plant[a]) Health Emergencies as Modified from USDA's *Draft Federal Emergency Response Plan for Outbreaks of Foot-and-Mouth Disease and Other Highly Contagious Diseases.*[b]

	USDA	FEMA	NCS	FCA	EPA	DOT	DOD	GSA	HHS	DOE	DOI	DOC	ARC	SBA	DOL	DOJ
Risk assessment	X	X						X	X		X					
Public education	X	X						X	X				X			
Surveillance, diagnosis, epidemiology	X				X		X	X	X		X					
Animal welfare	X							X	X		X					
Biosecurity	X							X	X							
Appraisal	X						X									
Indemnification	X															
Quarantine	X				X	X	X		X							
Vaccination	X							X	X							
Depopulation	X				X		X	X	X		X					
Disposal					X	X	X	X	X		X					
Cleaning, disinfection (site premises)					X		X	X	X							
Decontamination (equipment)					X	X	X	X	X							
Coordination, control (incident management)	X	X				X				X	X					
Interagency communication	X		X					X			X					
Public information, rumor control	X		X					X					X			
Notification, reporting	X	X	X					X		X						
GIS mapping	X		X					X			X					

Environmental monitoring, plume projections					X		X	X	X		X	X				
Permit issuance	X				X											
Long-term recovery	X	X		X										X		
Transportation						X	X	X								
Site security, perimeter control						X	X	X								
Law enforcement	X[c]							X			X					
Mass care for workers								X	X							
Public health, safety	X							X	X						X	
Mental-health counseling								X	X				X			
Intelligence							X[d]									X[d]
Other logistics	X	X				X	X	X			X					

[a]Although this table is taken from an emergency response plan for animals, it is the committee's sense that the response capabilities are also applicable to plants.
[b]Abbreviations for federal agencies and organizations are explained in Appendix C.
[c]USDA Office of the Inspector General (OIG) would probably be involved in law enforcement in USDA. The committee marked this role in the table, although it does not appear in the USDA plan.
[d]Not included in USDA's table because its plan addresses primarily unintentional occurrences, but the committee added these details because they would be relevant in the case of a suspected agricultural bioterrorism event.

SOURCE: Adapted and modified from US Department of Agriculture, Animal and Plant Health Inspection Service, Veterinary Services, 2001.

on collaborations with other groups, agencies, and individuals. Those entities include academe, the USDA Cooperative State Research Education and Extension Service (CSREES), professional societies, industry groups, other USDA units, other federal agencies, state departments of agriculture, state officials, and international organizations. Much of the nation's expertise and the laboratories designed to make critical diagnoses of agriculturally important pests and pathogens are in the universities and ARS. Professional scientific societies provide peer-reviewed reports of new pests and pathogens and may be the best sources of 1) databases on global distributions of agriculturally important pests and pathogens and 2) global authorities on their infestations and diseases. Industry groups are important—if not the only—sources of information on handling, shipment, and traceback of commodities. Industry groups are also one of the first-line responders to the appearance of a new pest or pathogen.

Likewise, the research and education programs needed to manage and recover from the appearance of a new pest or pathogen typically involve land-grant university and ARS programs (developed in cooperation with industry), state departments of agriculture, state officials, and APHIS. Although there are some efforts in this area (Lautner 2001), the committee did not find a concerted educational plan involving these groups. There is a need for APHIS—working in cooperation with cooperative extension, academic institutions, scientific professional societies, state animal and plant health agencies and industry representatives of US plant and animal commodities—to develop a comprehensive educational plan aimed at increasing the awareness of pests and pathogens new to US agriculture (Chapter 5, Finding I.H and Recommendation I.E.3.d).

Several USDA plans deal with unintentional introductions of plant and animal pests and pathogens, but the committee could not find any publicly available interagency or interdepartment national plan designed for response to an intentional introduction of a plant or animal pest or pathogen (Chapter 5, Finding I.B). However, the Terrorism Annex to the Federal Response Plan suggests that when requested by USDA or state and local governments, and within available resources, the Federal Emergency Management Agency (FEMA) would assist USDA by coordinating federal-agency support activities and delivering assistance to affected state and local governments. When it is suspected that an outbreak is the result of terrorist activity, the USDA Office of Inspector General (OIG) would notify the Department of Justice's Federal Bureau of Investigation (FBI). The OIG and FBI would jointly conduct the criminal investigation (see Table 2-1).

The roles and responsibilities of USDA, FEMA, other federal agencies, and state and local entities in responding to animal-disease outbreaks are delineated in the draft APHIS emergency response plan for animal disease (USDA-APHIS-VS 2001). Corresponding roles and responsibilities for plant emergencies are not as well defined (USDA-APHIS-PPQ 2002) and need further clarification (Chapter 5, Recommendation I.C). The FBI leadership role in responding to all intentional agricultural attacks also needs clarification (Chapter 5, Recommendation I.C).

If an outbreak cannot be adequately addressed under USDA or other federal authorities, the president may elect to declare an emergency under the Robert T. Stafford Disaster and Emergency Assistance Act (FEMA 1979). The act provides federal assistance to state and local governments to help alleviate suffering and damage from disasters.

DETERRENCE AND PREVENTON, DETECTION AND RESPONSE, RECOVERY AND MANAGEMENT

The goal of the US animal and plant health safeguarding system is to prevent the introduction and establishment of exotic animal and plant pests and pathogens or to mitigate their effects when eradication is not feasible. The United States uses an integrated network of intervention strategies targeted at key points in the system. Recent reviews have documented the need for major strengthening of the system to protect against unintentionally introduced pests and pathogens (National Plant Board 1999, NASDARF 2001).

The fundamental premise underlying the current system is that pest and pathogen introductions are unintentional and occur through natural migration across our land borders, artificial movement by international commerce (travelers, conveyances, cargo, and mail), or a combination of artificial and natural spread. Pre-entry, entry, and post-entry intervention strategies (preclearance programs, barrier zones, quarantines, inspections, detection, surveillance, and emergency response) are designed to mitigate unintentional events. However, recent emergency-preparedness exercises dealing with multiple introductions of FMD indicate that APHIS is beginning to address the possibility of multifocal intentional introductions (Iowa Department of Agriculture 2001, USDA-APHIS-VS 2000, USDA-APHIS 1999b).

Reliance on the US system to detect and respond quickly to an unintentional introduction may be an effective intervention strategy, but in most cases other actions are probably needed to strengthen the US animal and plant health safeguarding system against terrorist attacks. APHIS Plant Protection and Quarantine (PPQ) relies heavily on port-of-entry pest-interception data to identify potential foreign plant-pest threats. However, pests that are not likely to be introduced into the United States unintentionally could still constitute a potential threat if introduced intentionally. A system designed for unintentional threats is not sufficient for defending against intentional threats (Chapter 5, Findings I.B, I.H, and Recommendation I). The easy access in foreign countries to naturally occurring exotic pests and pathogens that can be intentionally introduced into the US requires a comprehensive reassessment of the strengths and weaknesses of the US animal and plant health safeguarding system.

Figure 2-1 depicts various points where intervention to deter, prevent, thwart,

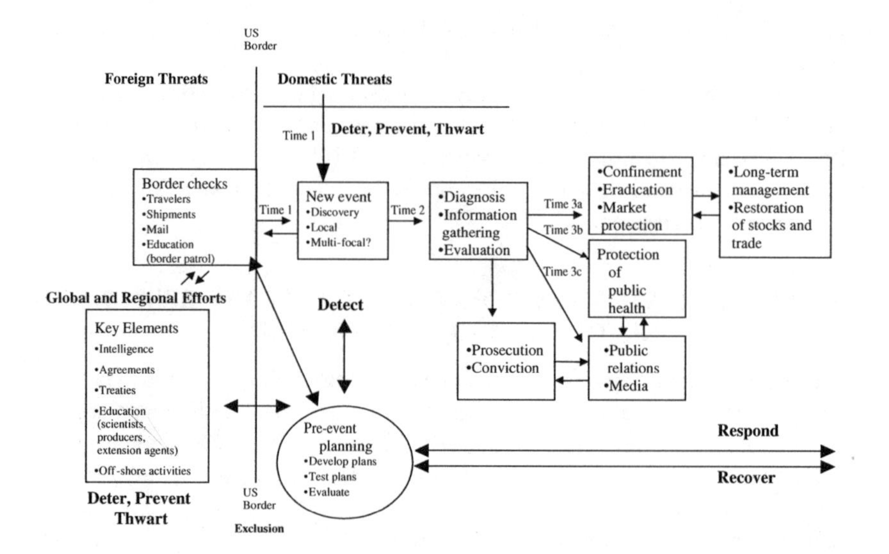

FIGURE 2-1 Event diagram and safeguarding intervention strategies. Diagrammatic outline of components, issues, and necessary preparation for and responses to a bioterrorist act directed at US agriculture. Various state and federal agencies are responsible for many components outside and inside the United States. Three elements of "time" are critical to detection, response, and recovery. Time 1 is time between release of infectious agent and its detection in crop, dairy, feedlot, storage facility, or other element of US agriculture. Time 2 is time required to make positive diagnosis or identification and begin to gather relevant information on scope and seriousness of disease or infestation outbreak. Time 3 has three components that must occur simultaneously: confine the problem, protect public health, and communicate with public.

detect, respond, and recover can take place. It is unlikely that intervention at any one point can provide acceptable security against intentional introductions; a series of interventions is more likely to reduce the threat of an introduction or mitigate its consequences after introduction (Chapter 5, Recommendation I.A). Current intervention activities and ideas for improving them are summarized in the sections below.

Deterrence and Prevention

Deterrence and prevention are considered the first lines of defense against the introduction of animal and plant pests and pathogens from foreign or domestic sources. Global and regional strategies are directed at reducing a potential threat before it reaches the US border, whereas the border strategy focuses efforts on interdicting a threat agent at US ports of entry. Smuggling-interdiction efforts can act as deterrents before biological agents reach their target. Domestic security initiatives prevent unauthorized releases by someone who gains access to the agent, for example, through US pest and pathogen research facilities.

Primary prevention and deterrence interventions for threat agents of foreign origin include international treaties and standards; bilateral and multilateral cooperative efforts; off-shore activities in host countries; port-of-entry inspection; quarantine; treatment; import testing; post-import quarantine; post-import handling of plants, plant products, animals, and animal products and byproducts; and smuggling interdiction. Civil and criminal penalties may be levied against violators of federal law involving these interventions.

Methods to prevent or deter threats of domestic origin are needed and could include increased surveillance of US laboratories or better on-farm security to prevent transfer of threat agents to the target (Chapter 5, Recommendation I.B). There are possibilities that pests and pathogens already established in the United States and confined to certain geographic areas could be transferred by terrorists to unaffected areas.

Global Strategy

The Office International des Epizooties (OIE) serves as the world animal health organization. OIE was founded in 1924 and has 158 member countries. Its main objectives are to inform governments of the occurrence and course of animal diseases throughout the world and of ways to control them; to coordinate, at the international level, studies devoted to the surveillance and control of animal diseases; and to harmonize regulations for trade in animals and animal products among member countries (OIE 2002).

The International Plant Protection Convention (IPPC), under the auspices of the United Nations, Food and Agriculture Organization (FAO), is a treaty dating back to 1952. The IPPC is aimed at promoting international cooperation to control and prevent the spread of harmful plant pests and pathogens associated with the movement of people and commodities; it provides the framework for the development and application of harmonized phytosanitary measures and the elaboration of international standards. Signatory countries—of which there are 116, including the United States—are required to cooperate in the exchange of information on occurrence, outbreak, or spread of pests and pathogens that may be of immediate or potential danger (IPPC 2002). The ability of countries to comply with the provisions of the convention varies greatly (Chapter 5, Recommendation I.B). Information available to APHIS concerning plant-pest and pathogen conditions in other countries is often sketchy at best and not always timely (Chapter 5, Finding I.C).

Regional Strategy

APHIS is actively involved in efforts to harmonize quarantines, exclusion strategies, and other safeguarding initiatives among countries that share our borders or that have similar plant and animal health conditions (National Plant Board 1999). Efforts to protect North America through a multi-national approach to plant-pest and pathogen exclusion are under way under the auspices of the North American Plant Protection Organization (NAPPO). NAPPO is affiliated with FAO and adheres to IPPC. NAPPO coordinates the efforts among Canada, the United States and Mexico to protect their plant resources from the entry, establishment, and spread of regulated plant pests and pathogens. Similarly, OIE uses regional organizations for many of the same purposes. The United States is an active member of the Region of Americas, which includes OIE member countries of both North America and South America; the United States, Canada, and Mexico constitute a formal tripartite group. Maintaining a high degree of security against unintentional introductions along the 7,000 mi of the US borders with Canada and Mexico is difficult at best. Additional cooperative efforts between Canada, Mexico, and the United States could help protect North America from intentional introductions (Chapter 5, Recommendation I.B).

Offshore activities are designed to mitigate pest and pathogen threats to the United States at points of origin and use various strategies. APHIS personnel are stationed in 27 host countries and are involved in surveillance and barrier programs, import and export trade facilitation, and commodity-preclearance programs. The largest number of APHIS employees stationed outside the United States is in Mexico. They work closely with Mexican agricultural officials to maintain pest- and pathogen-barrier programs to protect the United States, monitor pest and pathogen conditions in Mexico, and facilitate the import and export of agricultural products.

ARS operates six overseas locations for research on biological control of pests and pathogens. The research contributes to 1) accurate identification of foreign pest and pathogen species; 2) knowledge of basic biology; 3) habitat characterization; 4) assessment of ecological requirements; 5) knowledge of limiting environmental conditions and patterns of occurrence; 6) climate-matching; and 7) identification of potential control agents for foreign species (USDA-ARS 2002). Such information is essential to assessing pest and pathogen risk, developing rapid-detection tools, and planning responses (Chapter 5, Recommendation I.F.1.a).

The programs noted above target primarily unintentional threats, but intelligence-gathering domestically and in various regions around the world is vital for the appropriate deterrence and prevention of and preparation for intentional threats (Chapter 5, Recommendations I.B and I.E.1.a). Various US intelligence and law enforcement agencies collect information about programs that intend to develop or are developing biological weapons directed toward US agriculture (Murch

2001a). In recent years, a few USDA staff have been detailed to intelligence and law-enforcement organizations (Murch 2001a). However, it is unclear to the committee what information or approaches have been gleaned or whether findings have been incorporated or used.

Through those and other international efforts, scientists, extension agents, and other experts can be informed of each others' work and be more likely to notice aberrant research and development that might constitute a threat. By sharing information, ideas, and programs, international scientific collaboration can fortify national systems for safeguarding plants and animals from intentional threats (Chapter 5, Recommendation I.E.1.a).

Port-of-Entry and Smuggling Interdiction

In the United States, quarantine policies and procedures for foreign animal and plant pests and pathogens are in place to regulate the importation and in-transit movement of pests, pathogens, plants and plant parts, animals, and animal products byproducts. Systems for giving permits authorize the entry of regulated articles under prescribed conditions. Fifty conditions of animals are designated as foreign-animal diseases (FADs). There are 402 regulated plant pests (USDA-APHIS 2001f). Also, APHIS takes quarantine action on a large number of plant pests and pathogens on the basis of attributes of their particular taxa. Several lists of plant pests and pathogens potentially threatening to the United States have been developed with criteria that depend on the interests of the developers of the lists and the lists' intended uses. An interagency consensus threat list—developed by the intelligence, agriculture and science communities—seems not to exist. Such a list would be useful in planning, defining roles, allocating resources, and coordinating activities between agencies and sectors (Chapter 5, Recommendation I.D). It need not be exhaustive but should be representative of the various classes and types of agents deemed to pose the greatest threats to agriculture (according to likelihood of intentional introduction and likely impact). The choice should be based on the epidemiology and ecology of the threatening agents, potential economic effects, ease of containment, and ease of weaponization (Madden and Scherm 1999, Madden and van den Bosch 2002).

The effectiveness and the level of quarantine security achieved at ports of entry are directly related to the cooperation of the border Federal Inspection Service (FIS) agencies. The FIS agencies—namely the US Immigration and Naturalization Service (INS), US Customs Service (USC), and APHIS—are responsible for monitoring the entry of conveyances, passengers, cargo and mail at over 300 international ports of entry. INS and USC inspectors are cross-trained by APHIS to enforce animal and plant regulations at lower-risk ports of entry, perform initial screening of foreign travelers, and refer those of possible agricultural interest to APHIS inspectors at higher-risk ports of entry. APHIS staffs 126 higher-risk ports of entry with over 5,000 APHIS-PPQ officers, veterinary medical officers, and other personnel (Moseley 2001). The APHIS staff is supplemented by 62 specially trained agricultural canine teams who assist with international passenger baggage and mail inspection.

Port-of-entry inspection procedures include written and oral declarations, oral screening, canine screening, x-ray screening, and physical inspection. Those procedures, although they may be useful as a deterrent to intentional biological threats, are not specifically directed at these threats (Chapter 5, Finding I.D). There seem to be no short-term solutions, highlighting the need for novel inspection technologies and procedures to address the threat of intentional introductions (Chapter 5, Recommendations I.E.1.c and I.F.2.a).

In addition to the US Border Patrol and US customs agents, APHIS officers and state department of agriculture officers support smuggling-interdiction efforts at ports of entry and at nonport locations by intercepting contraband shipments after they have left the border. State highway patrol officers also may assist in apprehending violators. State highway patrol officers have been active in supporting agriculture inspections at highway weigh stations and other checkpoints. With the exception of federal and state agriculture officials, these agents and officers are generally not specifically trained to intercept agricultural violators, but they are an integral part of the smuggling-interdiction effort.

All travelers, conveyances, cargo, mail, and other articles arriving in the United States from foreign origins are subject to inspection, but only a relatively small percentage are actually inspected by APHIS (Table 2-2) (Chapter 5, Finding I.D).

The port-of-entry intervention strategy depends heavily on voluntary compliance by importers and the traveling public (Chapter 5, Finding I.D and Recommendation I.F.2.a). Foreign arriving travelers, conveyances, mail and cargo are selected for inspection and inspected at various rates based on the perceived pest

TABLE 2-2 Port-of-Entry Workload Data, Fiscal Year 2000

	No. Arrivals[a]
Passengers, crew, pedestrians, bags	129,054,465
Privately owned vehicles	128,602,841
Passengers in privately owned vehicles	332,834,888
Buses	460,408
Trucks	11,575,243
Maritime cargo bills of lading[b]	6,678,589
Mail packages	5,000,000
Farm animals	6,300,000

[a]SOURCE: US Customs data.

[b]Each commercial cargo shipment is given bill-of-lading number for manifest-tracking purposes. Bill can represent one or multiple units of cargo.

or pathogen threat. APHIS effectiveness data from fiscal year 2000 indicate that 95.2% of international air travelers complied with agriculture quarantine regulations, according to statistical sampling with a ± 0.5% margin of error (USDA 2000b). Data on the other pathways of entry are not available.

Current risk assessments are based on the premise that new pest and pathogen introductions are likely to be unintentional. Consideration of intentional introductions in these assessments may necessitate changes in inspection policies, priorities, and decisions. The following examples are provided to illustrate the need for specific attention to intentional pathways in the risk-assessment process (Chapter 5, Finding I.H and Recommendation I.E.2.a).

Traditionally, APHIS has considered movement across the US-Canadian border a low threat for the introduction of exotic pests and pathogens. There are 83 US Customs ports of entry along the Canadian border. Inspections conducted by APHIS over the Labor Day weekend in 2001 indicate that the pest and pathogen threats may be greater than believed. Intensive inspection of 4,000 vehicles crossing the US border at Alexandria Bay and Buffalo, NY, and Port Huron and Detroit, MI, resulted in the seizure of 6.5 tons of prohibited material and the interception of over 200 pests and pathogens that could threaten US agriculture (USDA-APHIS 2001a). It should be noted that although the data indicate potential risk, a qualitative analysis of the data is needed to describe the pest or pathogen risk accurately. Nevertheless, these data illustrate that the Canadian border might be vulnerable to both unintentional and intentional introductions of new animal and plant pests and pathogens. Other pathways that have been considered of low risk on the basis of unintentional introduction data include private aircraft and yachts, cruise ships, and express air shipments (National Plant Board 1999).

Foreign animal and plant pests and pathogens may be imported into the United States for identification and research when authorized by APHIS and with the concurrence of the destination state's animal- or plant-health authorities. Permits specify the restrictions placed on the handling and disposition of an organism and facilitate its movement through the port of entry to the final destination. Organisms are imported for research, biological-control projects, and taxonomic purposes. The permitting process depends heavily on the integrity of the applicants (Chapter 5, Finding I.D). APHIS recognizes those vulnerabilities and has temporarily suspended issuing new permits for some types of organisms under certain circumstances until critical deficiencies can be corrected (USDA-APHIS-PPQ 2001a). Strengthening current procedures could help to ensure that unauthorized exotic animal and plant pests and pathogens are not intentionally introduced into the United States by this pathway.

The United States imports about 700 million units of plant material for propagation each year. Seed representing some 90 species—including corn, wheat, and cotton—from over 100 countries is imported. Much of the seed is imported from off-season growing sites to accelerate varietal development. Admissible seeds are sampled and visually inspected at US ports of entry by APHIS. The visual inspection is aimed primarily at detection of foreign noxious weeds. No laboratory examination is done by APHIS for seedborne pathogens or the presence of unauthorized genetic material (Chapter 5, Recommendation I.F.2.a). The system relies heavily on internal controls by the seed-trade industry to prevent the introduction of exotic organisms (Chapter 5, Finding I.D).

Pollen is regulated by APHIS when it is imported as food for bees (USDA-APHIS 2001j). Pollen for human consumption or any other purpose is not regulated by APHIS.

The above examples suggest that assessments of the risk associated with various pathways of intentional introduction are needed (Chapter 5, Recommendation I.E.2.a).

Violators of the Plant Protection Act (PPA) and animal-health statutes are subject to civil and criminal penalties. The secretary of agriculture has the authority to assess civil penalties up to $50,000 for individuals and $250,000 for businesses that violate the PPA. However, the law limits the penalty to $1,000 for first-time offenders carrying an illegal agricultural product through ports of entry for personal use (US Congress 2000). Similar authority is included in the Animal Health Protection Act pending before Congress (US Congress 2001). Criminal prosecutions are pursued only in the most egregious cases when individuals or other entities knowingly violate the laws.

Domestic Security Programs

The United States does not have a formal program for preventing agricultural threats that might arise domestically (Chapter 5, Finding I.D and Recommendation I.B). For example, a perpetrator might gain access to a foreign disease agent through a research laboratory and target US livestock or crops. Two kinds of security need to be considered: laboratory security to prevent access to the threat agent in the first place and on-farm security to prevent introduction to the target. Recent proposed legislation would increase surveillance and security of laboratories working with dangerous pathogens (e.g., HR 3160, the Bioterrorism Prevention Act). It is difficult to envision how introduction of a threat at the farm level might be prevented, given the necessity to raise plants and animals in relatively open and diffuse spaces. However, farmers could be made more aware of the possibility of intentional threats and increase their vigilance, and local law enforcement officials could be better prepared to recognize suspect activity and be stationed at large trade shows or auctions.

Detection and Response

Detection and response activities are designed to react to any breach of the prevention effort. Procedures for initiating and conducting response activities are outlined in the PPQ draft *Emergency Programs Manual* (USDA-APHIS-PPQ

2002) and in the Veterinary Services (VS) draft *Federal Emergency Response Plan for an Outbreak of Foot-and-Mouth Disease or Other Highly Contagious Diseases* (USDA-APHIS-VS 2001). The basic approaches and the types of resources needed for rapid detection and response to animal and plant pests and pathogens are similar, but the wide disparity in the number of potential threats, the level of preparedness, and other factors warrant separate discussions.

Animal Pests and Pathogens

Detection of foreign-animal pests and pathogens and the resulting infestation and disease starts with initial observations by livestock producers and their private veterinary practitioners (Figure 2-2). Some 40,000 private-practice veterinarians are accredited by APHIS; this means that they have some responsibility for the identification of FADs. However, the majority of the veterinarians have not had training in the recognition or response to FADs since receiving their initial accreditation. Some 450 public veterinary practitioners, most of them state and federal employees, have received special hands-on training in FADs and are strategically disposed around the 50 states (USDA-APHIS 2001k). They are designated foreign-animal disease diagnosticians (FADDs) and are able to be on site at suspicious premises within 8 hours in most cases and within 24 hours in all cases. Definitive diagnoses are made at the APHIS National Veterinary Services Laboratories in Plum Island, NY, or Ames, IA.

Laboratory-based domestic surveillance systems are in place for 22 of the most important livestock and poultry diseases, most of which are FADs. Surveillance is most aggressive for diseases that have recently been eradicated or are in the final stages of eradication, such as hog cholera, brucellosis, and highly pathogenic avian influenza. In addition, APHIS conducts regular national animal-health monitoring surveys, each evaluating a variety of diseases in a given commodity area.

Current programs rely on initially detecting the presence of pests and pathogens by observing the health of animals (Chapter 5, Finding I.D). However, the lag from initial infection to display of clinical signs ranges from 2-14 days for FMD to years for BSE.

Initial response to the diagnosis of an FAD is at the local and state level. The major livestock-producing states have emergency response plans in place and organizations and varying levels of resources to carry out those plans. Other states presently have limited resources for mounting a significant response. If an outbreak spreads across state lines or if local and state efforts are unable to control the outbreak, federal involvement quickly follows. The core of a federally coordinated field response is carried out by a Regional Emergency Animal-Disease Eradication Organization (READEO). Each of the two regions in the United States (western and eastern) has a READEO consisting of 38 team members available for immediate call-up in the event of a serious outbreak. The

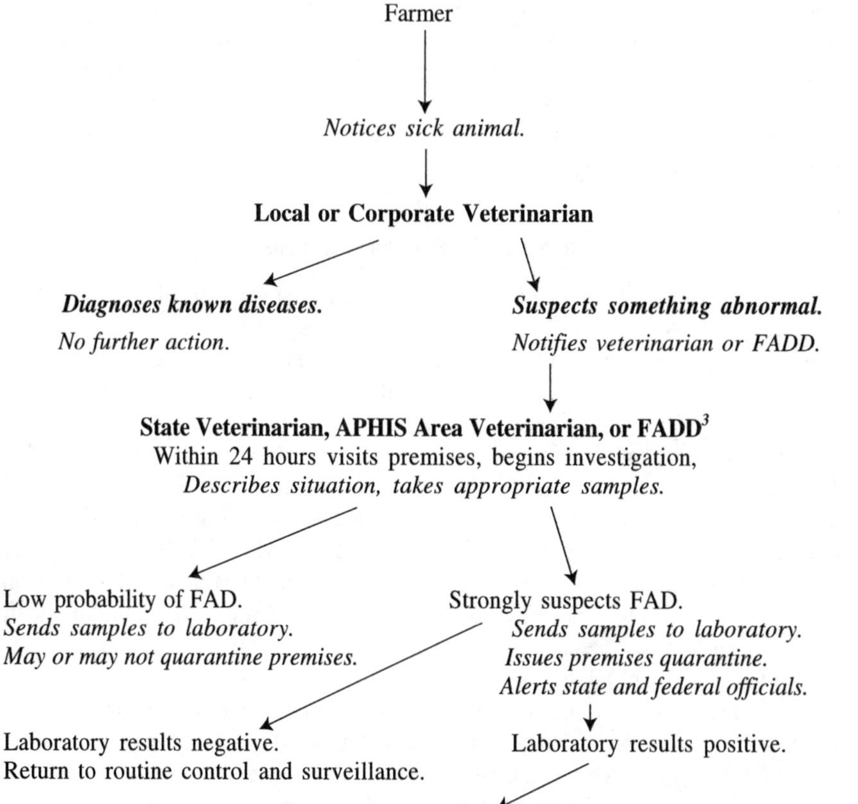

Emergency Response Team (ERT)

Describes situation. Begins epidemiological investigation and traceback. May expand quarantine.

Regional Emergency Animal Disease
Eradication Organization (READEO)

Coordinates national communication; manages national consequences and federal response; starts eradication; coordinates with **Federal Emergency Management Agency** according to Federal Response Plan (USDA-APHIS-VS 2001).

FIGURE 2-2 APHIS emergency procedures for animal disease outbreaks. SOURCES: US Department of Agriculture, Animal and Plant Health Inspection Service, Veterinary Services 2001 and Kohnen 2000.

[3]The FADD may be a state or federal veterinary medical officer.

teams are well trained and engage in periodic test exercises based on simulated outbreaks. A recent test exercise involved a simulated release of FMD virus in three countries (Chapter 5, Recommendation I.E.2.a) (CFIA 2000). The broad-based response included federal, state, and industry personnel. Deficiencies identified included communication-systems problems (Chapter 5, Recommendation I.E.1.d), both within and among the countries, and uncertainties about the application of a decision matrix for determining the use of a vaccine (Chapter 5, Recommendation I.E.2.b).

An important planning initiative, the National Animal Health Emergency Management System seeks to have in operation by 2005 a comprehensive emergency-management system that integrates the efforts of federal agencies, state agencies, the animal industries, and private veterinary practitioners focused on prevention, preparedness, response, and recovery. The effort is being led by representatives of four national groups: the American Veterinary Medical Association, the Animal Agriculture Coalition, APHIS, and the United States Animal Health Association. At the state and local levels, efforts are under way in all 50 states to develop animal-health emergency-management plans. Standards for state animal-health emergency-management systems were developed to evaluate state emergency-response preparedness. States must have in place emergency management plans, written agreements specifying roles and responsibilities, authorities and policies, surveillance, communication, training and education, and funding. The states are in various stages of compliance with the standards (USAHA 2002).

Specific protocols for response to many of the most serious FADs have been published by APHIS in a series of documents called redbooks. These are now being replaced with a more generic series of response documents in the comprehensive *Federal Emergency Response Plan* (USDA-APHIS-VS 2001).

A unique feature of the response plan for FMD consists of scenarios in which vaccination would and would not be used. A North American FMD vaccine bank has been established jointly by the governments of the United States, Canada, and Mexico to provide vaccine with a minimal waiting period.

The APHIS animal-health safeguarding system has had substantial success in controlling and eradicating endemic animal diseases, such as hog cholera, tuberculosis, and brucellosis. Control methods vary depending on the agents and have been limited largely to identification, quarantine, rapid slaughter, and disposal of affected animals. Most FAD agents that were unintentionally introduced from outside the United States have been eliminated. For example, the last outbreak of FMD was eradicated from California in 1929. Since then, introductions of velogenic viscerotropic Newcastle disease (1972) and Venezuelan equine encephalitis (1971) and several relatively small outbreaks of vesicular stomatitis in recent years have been eliminated. Along with the federal government, state governments, industry, and a number of allied groups have been involved in these efforts. Current animal-disease eradication programs include those for cattle and swine brucellosis, bovine tuberculosis, and pseudorabies in swine. A notable exception to eradication success is the recurrent outbreaks of West Nile Virus in recent years.

In 2001, the National Association of State Departments of Agriculture concluded its *Animal Health Safeguarding Review* (NASDARF 2001). The focus of the review was limited largely to existing systems for preventing and responding to serious incursions of animal diseases. Over 150 recommendations were made to strengthen the current system. Although this review provides valuable recommendations that could be utilized in a plan to counter bioterrorism, little specific attention was given to intentional introductions (Chapter 5, Finding I.B). Clearly, multiple incursions of an FAD, such as might be the case in a planned act of bioterrorism, could quickly overwhelm currently available state and federal resources. Plans are being made for making other human resources available, such as accredited private veterinarians and retired federal employees.

Plant Pests and Pathogens

Rapid detection and diagnosis are critical for eradicating or mitigating the effects of newly introduced plant pests or pathogens (Figure 2-3). The lags between introduction, discovery, and detection of a plant pest or pathogen are difficult to determine and vary widely, depending on the nature of the organism and the degree of surveillance (Chapter 5, Finding I.D).

A 1999 review of the plant-health safeguarding system (National Plant Board 1999) describes current efforts to detect the introduction of exotic plant pests and pathogens, including passive detection and active surveillance. Passive detection occurs during other scientific field work, such as crop surveys, population studies, fauna or biodiversity surveys, endemic species inventories, and incidental reports from growers and the general public. Active surveillance documents the presence or absence of pest or pathogen species targeted by specific survey programs. Detection methods are usually species-specific as determined by pest or pathogen biology and the availability of a survey tool. Federal survey guidelines are available for 16 species of insects (USDA-APHIS-PPQ 1991a,b), including several species of fruit fly and the Asian longhorn beetle, Karnal bunt, and golden nematode.

PPQ administers a federal-state cooperative agricultural pest survey program. It was created to provide a national plant-pest and pathogen detection network by combining federal and state resources for surveys. State contributions are inconsistent from state to state and region to region (National Plant Board 1999). Survey data are accessible through the National Agricultural Pest Information System. The program targets invasive plant pests and pathogens, plant pests and pathogens affecting export, and domestic federal-state cooperative pest-management programs—such as those for Karnal bunt, gypsy moth, and imported red fire ant—pests not widely geographically distributed in the United States. Individual states often survey for pests and pathogens that are not covered by federal

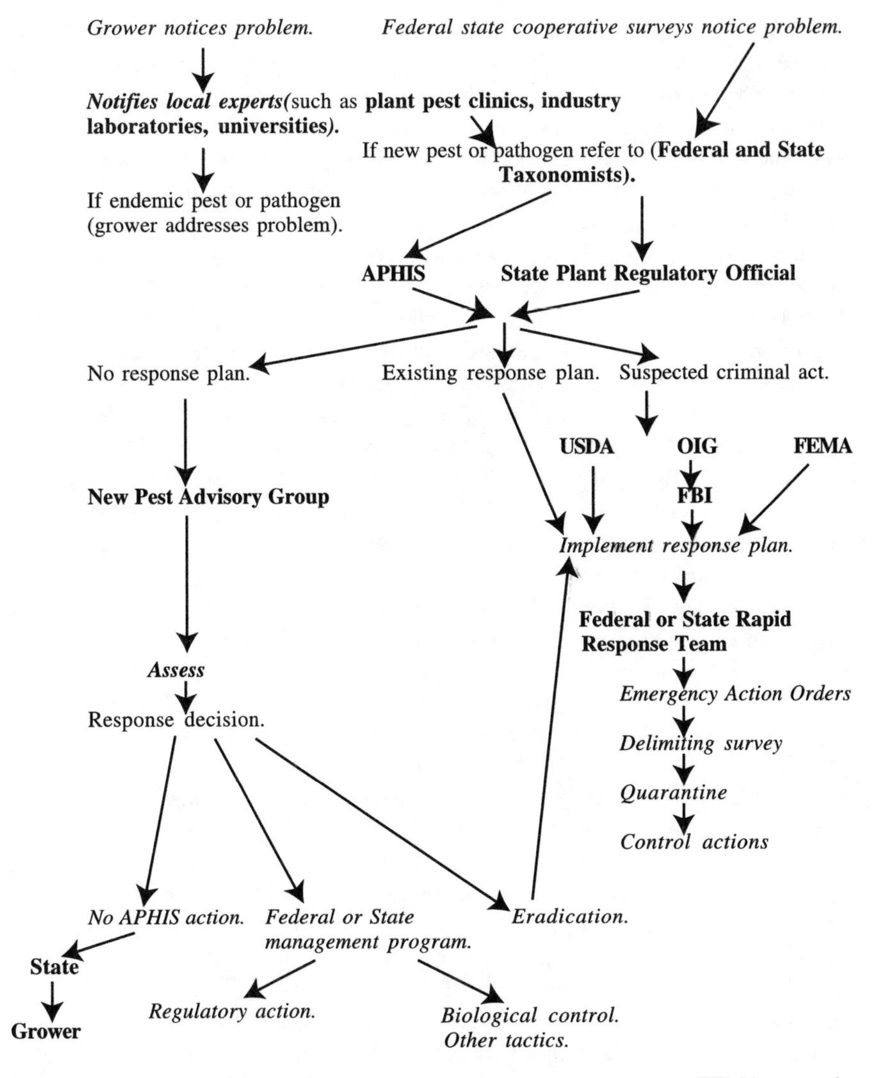

FIGURE 2-3 APHIS plant pest or pathogen emergency response system. FEMA responds when requested. Advisory group gives response decision. SOURCES: USDA-APHIS-PPQ 2002, Kohnen 2000.

programs but are of interest to their agricultural producers. Survey, identification, and reporting procedures are defined in the APHIS exotic pest survey manuals.

No clearly defined system exists for screening, identifying, and reporting to PPQ plant pests and pathogens detected by ad hoc surveillance (Chapter 5,

Recommendation I.E.1.a and b). Current identification and reporting procedures rely heavily on the knowledge, awareness, and voluntary action of the involved parties. Organisms found by growers, county extension agents, university scientists, state and federal agencies, or other interested parties can be submitted to PPQ, the ARS Systematic Entomology Laboratory, or directly to an appropriate academic scientist for identification. In practice, these organisms are usually first turned into industry laboratories (such as commercial seed company laboratories) or shown to crop consultants or county agriculture agents. When suspicious organisms or symptoms are identified, they often are sent to university clinics for confirmation. University plant clinics play very important roles in early detection of arthropod pests, plant parasitic nematodes, and plant pathogens, but they typically are understaffed and lack resources. There is seldom time or resources to perform molecular or biochemical assays. The personnel in the laboratories might not be well versed in identification of exotic organisms. In addition, these laboratories commonly handle hundreds, if not thousands, of specimens each year from home gardeners; this creates the potential for backlogs in responding to growers' and agribusinesses' request for diagnoses.

Correct and timely identification or diagnosis is critical for the response decision. Delays in examining an organism or lack of awareness of foreign organisms can lead to further reporting delays and misdiagnoses, which can affect the timeliness of an emergency response for a new introduction. Furthermore, organisms may come to PPQ's attention several years after initial detection. Plum pox was discovered by a grower and turned in for identification to a university plant clinic in 1997 and again in 1998, and it was misdiagnosed each time. In 1999, the grower took samples to an industry fruit show, and an extension agent familiar with plum pox became suspicious. Additional diagnostic tests confirmed its presence 3 years after the initial detection.

Lack of 1) early-detection capability, 2) a universal plant-pest and pathogen identification and reporting system, 3) a supporting database, and 4) taxonomists trained to identify foreign organisms can severely handicap the response to new introductions. A comprehensive plant-pest surveillance system that takes full advantage of both passive and active surveillance activities, new technologies, and shared databases is needed to combat both intentional and unintentional introductions effectively (Chapter 5, Recommendation I.E.1).

APHIS has developed species-specific emergency-response plans for 28 high-priority exotic plant pests and pathogens. The response plans are activated and response personnel can be on site within 48 hours of confirmation of the pest or pathogen. In recent years, APHIS has relied heavily on the New Pest Advisory Group (NPAG) process with the capability to develop a response plan within 14 days of detection of a new pest or pathogen. NPAG is composed of scientific and regulatory experts from all over the world. The number of new or re-emerging potential plant pests and pathogens detected and reported to NPAG was 27 in 1998, 16 in 1999, and 58 in 2000 (USDA-APHIS-PPQ 2000). This approach has merit for the vast array of *unintentional* plant-pest or pathogen threats, but a more active response program is needed to ensure a quick response to *intentional* introductions. A consensus list, developed by the agricultural and intelligence communities, of the top biological pests and pathogen threats is needed to focus emergency planning and test emergency-response capability. Also, an intentional introduction may be of an endemic pest or pathogen (perhaps just a more virulent strain), and the APHIS plans would not even be implemented until it was established that the agent was a new strain (Chapter 5, Finding I.B and Recommendation I.D).

PPQ regional Rapid Response Teams (RRTs) are available when a federal response is needed (Figure 2-3). During January – September 2001, PPQ was involved in six federal-state cooperative eradication programs. An RRT is a cadre of PPQ program specialists identified and trained by one of the two PPQ regions to respond to plant-pest or pathogen emergencies within 48 hours under the direction of the PPQ regional director. The cadre is generally used to supplement the permanent workforce stationed in the area and to provide specialized startup expertise until permanent staffing can be put in place. The RRTs may be supplemented by available trained state personnel. The ability to provide emergency-response personnel and personnel from other federal agencies varies widely from state to state, depending on the agriculture infrastructure. Additional RRT staffing would probably be needed to respond to multifocal attacks, and national coordination would be needed for efficient use of the teams (Chapter 5, Finding I.F).

A 1999 review, *Safeguarding American Plant Resources,* conducted by the National Plant Board concluded that the current exclusion, detection, and emergency-response components of the plant-health safeguarding system can, with modifications, meet the demands of the global marketplace (National Plant Board 1999). The report contains more than 300 recommendations to strengthen the current system. Many of the recommendations—such as strengthening the APHIS biological knowledge base for conducting pest and pathogen risk assessments, deploying new inspection technologies at ports of entry, redesigning the pest- and pathogen-detection system, and providing a greater emergency-response capability—are applicable to both unintentional and intentional introductions. However, the review did not specifically address vulnerabilities in the system due to emerging pathways, such as bioterrorism and genetically modified organisms. Also, the review leaves unanswered the question of what, if anything, should be done differently to mitigate the threat of intentional introductions (Chapter 5, Finding I.B.1 and Recommendation I).

Recovery and Management

Consequence management and recovery are important elements of the overall safeguarding system for unintentional and intentional threats. As currently designed, programs focus on unintentional threats (Chapter 5, Finding I.B). However, there are differences in the steps needed for intentional threats, such as enhanced intelligence, forensics, and law enforcement (Chapter 5, Finding I.A). Several activities—such as confinement, eradication, market protection, public-health protection, communication, and law enforcement—start in the response phase but continue throughout the management and recovery phase (Figure 2-1). Elements of long-term economic recovery include replacing germplasm, rebuilding confidence in markets, and regaining international market share.

Those elements are outlined in Figure 2-4, which includes major steps in economic recovery from a bioterrorist attack on agriculture. The confidence of industry, as well as of domestic and foreign customers, depends on continuing surveillance after the threat agent has been eradicated or controlled. Protection of or plans for regenerating germplasm from the most productive or critical genetic lines is essential for full recovery. FEMA appears to be the lead federal agency during recovery (PDD 1995) and is authorized to provide assistance as requested by state and local governments. If financial assistance is granted, broad support to the sector would come through the USDA Farm Services Agency (FSA) because of its existing infrastructure for supporting the agricultural economy. If appropriated, financial assistance to the public infrastructure or individuals would come through FEMA (USDA-APHIS-VS 2001).

Full recovery may depend on eradication of the agent from the United States. If eradication is not possible, a new lower equilibrium with endemic disease or infestation would be established. The new equilibrium would be characterized

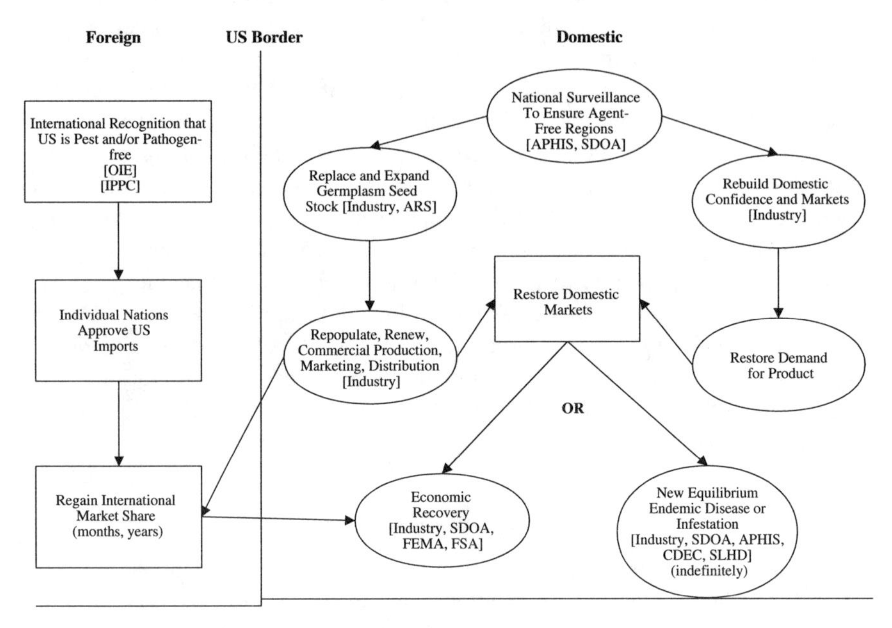

FIGURE 2-4 Economic recovery. Abbreviations are defined in Appendix C.

by 1) continuing losses due to disease or infestation; 2) costs of management and control, such as vaccination or development of resistant varieties of the affected crop; 3) continuing loss of some markets; and, in the case of a human pathogen, 4) continuing human health costs.

The APHIS plant-safeguarding system involves both emergency eradication protocols and long-term pest-management programs to mitigate the effects of newly introduced plant pests and pathogens when eradication is not likely. APHIS involvement in long-term cooperative management programs is limited to pests and pathogens not widely distributed in the United States, where biological control is the intervention of choice, or plant pests and pathogens as specified in the PPA of 2000. Long-term federal pest-management programs are conducted in cooperation with state departments of agriculture and producers under the auspices of a cooperative agreement that delineates the cost-sharing between the parties involved. A number of other federal agencies also address non-eradication responses—US Forest Service, ARS, CSREES, the Department of the Interior (for example, the Bureau of Reclamation), as do various state agencies and universities.

The APHIS animal-safeguarding system targets 50 foreign-animal disease conditions with eradication as the intended outcome. The USDA draft ***Federal Emergency Response Plan for Outbreaks of Foot and Mouth Disease and other Highly Contagious Diseases*** provides the most comprehensive framework available for responding to animal diseases (USDA-APHIS-VS 2001).

Federal and state recovery plans also need to address economic, social, and psychological consequences. Communication and education programs need to be in place to inform parties directly affected by an outbreak and, if consumers are affected, to maintain their confidence in the US ability to respond effectively to the outbreak (Chapter 5, Recommendation I.E.3). Restocking destroyed animals, implementing production-level pest- and pathogen-control programs, replacing susceptible varieties with resistant varieties, and restoring export markets are integral parts of economic recovery (Figure 2-4).

Federal Compensation

The PPA authorizes the secretary of agriculture to compensate any person for economic losses incurred as a result of action taken by the secretary when an extraordinary emergency is declared. The determination by the Secretary to pay compensation and the amount of compensation are discretionary, final, and not subject to judicial review. The US Code (US Congress 1962) requires the secretary of agriculture to pay fair market value for animals, articles, and materials destroyed by USDA. The Animal Health Protection Act was introduced in the US Congress in May, 2001. If passed, it would consolidate and update several existing animal-health laws (US Congress 2001a). Funds may come from reprogramming or contingency funds within USDA, transfer from the Commodity

Credit Corporation (CCC), or direct appropriations from Congress. If a special appropriation is made by Congress for compensation, a declaration of extraordinary emergency is not necessary. FSA handles compensation payments authorized by the secretary. From 1972 to 2001, $1.15 billion was transferred from the CCC for plant and animal eradication programs, of which $313.8 million was made available for compensation to individuals as a result of actions taken by the department (USDA-APHIS 2001b).

FEMA statutory authority (FEMA 1979) defines emergency as "any occasion or instance for which, in the determination of the President, Federal assistance is needed to supplement State and local efforts and capabilities to save lives and to protect property and public health and safety, or to lessen or avert the threat of a catastrophe in any part of the United States." The Small Business Administration (SBA) can provide economic-disaster loans to small businesses that are agriculturally dependent; however, SBA cannot provide loans to farmers.

ANIMAL AND PLANT HEALTH-SYSTEM FINDINGS

The committee recognizes that the US defense system for agriculture is evolving. Recently, the evolution has been rapid, given heightened concerns about bioterrorism (for example, after the anthrax attacks through US mail in fall 2001) and outbreaks of foreign disease agents (for example, after the FMD outbreak in the UK in spring 2001). Therefore, it is possible that the committee's concerns and suggestions for improving the system are already being addressed, without its knowledge, in the federal agencies.

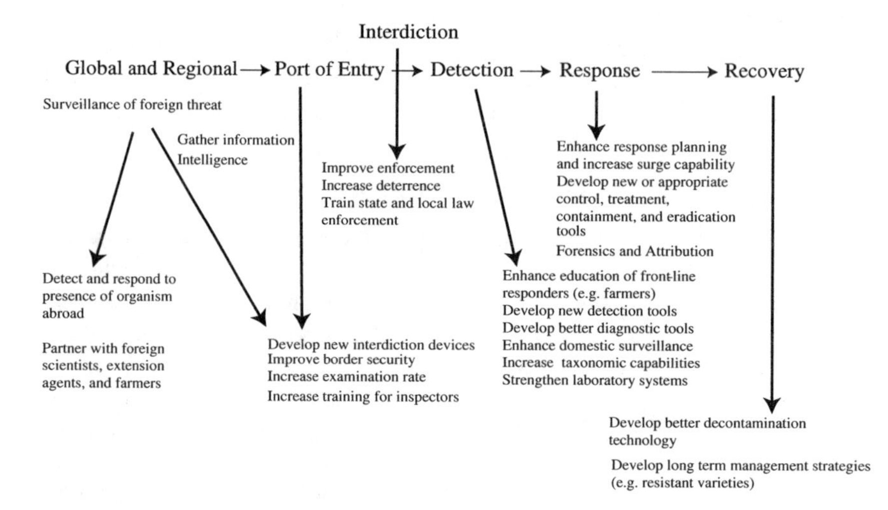

FIGURE 2-5 Strengthening the agricultural pest and pathogen safeguarding system.

In this chapter, the committee has analyzed the current system to protect US agriculture. Although it is strong in many respects, there is a need for improvement, especially in light of the plausibility of intentional threats. Figure 2-5 summarizes points in the system that could be strengthened. The committee's key findings, conclusions, and recommendations pertaining to the abilities of the system to deter, prevent, thwart, detect, respond to, and recover from threats appear in Chapter 5.

3

Lessons Learned from the Study of Potential Agents of Agricultural Bioterrorism

INTRODUCTION

Several potential agricultural bioterrorist agents were subjects of in-depth study by the committee in its assessment of the scientific, technical, and organizational strengths and weaknesses of the US defense system for bioterrorism directed at agricultural plants and animals. The committee drafted a detailed description of current knowledge of each agent as it relates to potential agricultural bioterrorism. This detailed information is published as an appendix that is not for distribution to the general public. General lessons learned from the committee's deliberations and in-depth analyses of these agents are presented in this chapter.

BASIS FOR SELECTION OF AGENTS STUDIED

The committee chose to examine specific agents so that it could address a variety of issues considered important in coping with agricultural bioterrorism. The issues include the following:

- Economic and public-health impacts.
- Research needs.
- Detection, diagnosis, and surveillance.
- Technology transfer.
- Training.
- Law enforcement.
- Global surveillance.

- Crop monoculture and industrial-scale production.
- Circumvention of host resistance to threat agents.
- Vaccination policy.
- Border control.
- Emerging or new threats.
- Potential for eradication.
- Trade embargoes.
- Leadership and coordination among sectors and agencies.
- Public information and credibility.
- Terrorist profiles and motivation.
- Intelligence.
- Biosecurity (farm, laboratory, and inputs).

The committee examined potential bioterrorist agents that could be considered to pose major as well as moderate or lesser threats in order to assess the capabilities of the US defense system for agriculture. The committee did not analyze animal welfare issues (for example, suffering, debility and death), but recognizes these as important factors in responding to bioterrorist threats.

The agents selected for study by the committee were not intended to be comprehensive with regard to types of threat agents or targets. The issues raised, as well as many of the strengths and weaknesses of the system as regards each agent examined by the committee, are representative of those applicable to a broad class of threats by similar agents or to similar targets. Although some sectors of agriculture—such as forestry, aquaculture, and poultry production—were not studied in detail by the committee, they are vulnerable to bioterrorist attacks, and their associated threats and issues are similar to those presented by the agents that the committee examined.

CHARACTERISTICS OF AGENTS STUDIED

The agents studied by the committee were not necessarily the most likely or potentially damaging bioterrorist threats to US agriculture. It was beyond the scope of the committee's charge to make that determination. The agents studied by the committee can be broadly categorized as: animal diseases and their vectors, contaminants, invasive insect pests, and plant pathogens and diseases.

LESSONS LEARNED

Lessons learned by the committee's in-depth study reflect the committee's consensus judgment and contain a mixture of findings, conclusions and recommendations. Many of these lessons are explicitly derived from the committee's review of specific agents, however others reflect the committee's larger understanding of and experience with the US defense system for agriculture as discussed in Chapter 2. In many instances, the lessons learned are not limited to a particular threat and target combination; rather, they are applicable to the larger system for defense against broad classes of plant and animal pests and pathogens.

Animal Diseases and their Vectors

The committee concludes the following from its in-depth study of several animal diseases and animal disease vectors that could be used as agents of agricultural bioterrorism.

1. Some animal diseases are of greater economic than public-health importance. Even though significant public health impacts could in principle result from intentional introduction of animal diseases, their spread would likely be minimized by regulatory bans and procedures.
2. Limitations of current diagnostic tests and current understanding of the pathogenesis and epidemiology of specific animal diseases may make these diseases suitable for use in hoaxes.
3. Regulatory controls can substantially reduce the likelihood of natural introduction of some animal diseases.
4. Because some animal diseases are not highly contagious, selective culling would be possible if we had sensitive on-the-hoof preclinical diagnostic tools.
5. Development of effective diagnostic and identification tools for the animal diseases of concern warrants a high research priority today.
6. Basic-science and technology programs will have broad application in protecting us from harm.
7. Effective public-information materials should be drafted in advance of natural or terrorist introduction of animal diseases of concern into the United States, so that they will be available immediately whenever needed.
8. The threat of an animal disease as an agricultural terrorist agent will be limited by factors such as

- Difficulty in obtaining or producing the agent.
- Physical and biological security of the plants that manufacture animal feeds, animal medicinals, and vaccines.
- Regulatory actions.
- An active national surveillance program.

9. Vulnerability to animal disease as an agricultural terrorist threat agent is increased by

- Limited effectiveness of border controls (for example, inspection procedures that are not developed with terrorists in mind, with a small proportion of luggage inspected at ports of entry).

- The small number and low sensitivity of diagnostic tests to detect an agent in living animals or animal tissues.
- A high resistance of an agent to inactivation by physical and chemical treatments.
- Lack of full compliance with regulations in place to control or eradicate the disease.
- A long incubation period from exposure to onset of disease, which would allow time for terrorists to escape detection and for wide dissemination of infected animals before discovery.
- An unwarranted degree of public concern over the disease, which could leverage a small number of cases or a hoax into an event with major adverse economic, social, and political effects.

10. Modern molecular field tests for animal diseases of concern need to be validated and introduced by USDA regionally and encouraged locally.

11. Vaccine stocks for animal diseases of concern need to be modernized and expanded.

12. Research should be performed to develop vaccines suitable for specific disease subtypes.

13. The United States should investigate the global eradication of those animal diseases posing significant threats and cooperate with international agricultural and wildlife experts in doing so. A continuing international mechanism to identify measures needed for global eradication of particular diseases should be established. Through such a mechanism, a global vaccination and eradication strategy could be developed with the participation of diverse experts and stakeholders. This could be a win-win situation for the United States and for other countries.

14. Widespread distribution of potential vector species increases the potential public health and economic impacts of a zoonotic disease.

15. It is essential for an effective response to have in place an infrastructure of disease surveillance and response systems, as well as cooperation and communication among agricultural, wildlife, and public health organizations.

16. Early detection and diagnostic tools are pivotal for limiting the extent of an outbreak. Education, limitation of animal movement, and development of vector population control methods are other important factors.

17. Basic research is critical for understanding of the pathogenesis and epidemiology of many animal diseases.

Contaminants

From its detailed examination, the committee finds the following as regards potential contaminants of seeds and agricultural commodities.

1. An introduced seedborne pathogen, if part of a national or multinational quarantine, or a single transgene, if unapproved, can have severe economic consequences, if found in a US seed supply or commodity intended for interstate shipment and export. These consequences can extend to the producers of the commodity, to rural communities and local agribusinesses supported by the commodity, and to the seed company and seed handlers. There can also be consequences for government agencies and taxpayers, through compensation for losses and expenses incurred because of the infestation or contamination. Those consequences can occur even in the absence of evidence that the pathogen or transgene is harmful to human, animal, or plant health.

2. Complete elimination of a genetic contaminant from an outcrossing crop could take a substantial amount of time. Likewise, it can take time to eliminate a seedborne pathogen from breeding material.

3. The economic damage caused by introduction of a single quarantined pathogen or illegal transgene could extend to food processors and to the export of the food commodity and not be confined to the seed supply.

4. The methods and laboratories available for diagnosis or testing for either a regulated seedborne pathogen or an unwanted transgene in seed or commodities are too few, too inefficient, or too slow relative for the needs of US seed and commodity industries (Chapter 2).

5. Many seedborne pathogens could be and should be deregulated when, on the basis of science or new technology, they become inconsequential to plant, animal, or human health or to the environment. In event of an intentional attack using such pathogens, deregulation could help farmers in the affected areas regain a position in the marketplace. There should be a mechanism and plan for 1) studying the importance of each pathogen with respect to political, economic, and scientific concerns and 2) deregulating the pathogen if it is not likely to be a major problem.

6. Pre-emptive breeding for resistance—that is, development of genetic resistance to plant pathogens that are not present but are potentially destructive to economically important crops—may need greater consideration in the planning and priority-setting process for USDA-, state- and industry-funded agricultural research and development. Such a strategy would be similar to the vaccination of people or livestock in anticipation of a contagious disease.

7. Confinement through quarantine by counties or states, although potentially economically harmful to the quarantined areas, can protect international markets for the unaffected areas of the United States. The effectiveness of that approach depends on observant and conscientious producers, farm workers, and commodity handlers at the local level who must report the regulated disease or infestation, a well as on a rapid diagnosis and response by the authorities representing state departments of agriculture and APHIS.

8. Educational campaigns are needed on potentially destructive pathogens and pests to assure their recognition at the local level.

Invasive Insect Pests

1. The expected result of a bioterrorism event involving a specific invasive pest could be an expensive but ultimately successful eradication program. However, the tools needed to successfully eradicate some invasive species might need to be developed.

2. Tools for responding to bioterrorism involving insect agents are lacking. Effective traps or other detection methods, federal and state action plans, eradication plans tested in the region of the insects' origins, pesticides registered for use against insect agents, educational plans for the agricultural community and general public, and legal understanding and enforcement to institute control measures are needed.

Plant Pathogens and Diseases

The committee's conclusions as a result of its in-depth examination of plant pathogens and diseases are presented below.

1. The large acreage of field crops is an important factor to consider regarding the introduction of plant pathogens.

2. Early thwarting of an attack against major food crops could be difficult because of the potentially long lags between introduction and first detection.

3. Use of resistant varieties is a standard way of controlling many important foliar diseases of field crops, but it is not practical or affordable to breed resistant varieties for every possible threat. For likely threat agents, resistant genotypes and molecular markers should be identified in advance, so that breeding efforts can proceed more quickly after an agent becomes established (Chapter 4).

4. Some plant pathogens and diseases are relatively easy to diagnose but diagnosticians may or may not be familiar with them. Increased efforts are needed to educate plant- disease diagnosticians in land-grant universities and state departments of agriculture to recognize a broader array of diseases, especially those which do not normally occur in a given region (Chapter 5, Recommendation I.E.3.d).

5. In light of potential disease reporting delays, diagnosticians and agricultural extension agents need to be better informed of USDA recommendations regarding the rapid reporting of unusual disease outbreaks. A Web-based interactive database is needed for routine reporting of such outbreaks; this may help in earlier accurate detection of deliberate introductions of plant pathogens (Chapter 5, Recommendation I.E.1). Although land-grant universities and state and federal agencies are important places for detection, it is possible that the first detection would be by a private industry, such as a seed company. Companies should be educated on the need to report unusual disease outbreaks to state and federal agencies rapidly. Furthermore, a well-advertised list of laboratories that

can conduct rapid molecular characterizations of major groups of pathogens should be prepared and distributed.

6. Strategies for recovering from deliberate plant pathogen attacks should be pursued. Those should be based on the epidemiology of the disease and will likely change depending on the particular pathogen and crop types.

7. Some plant diseases are difficult to control or manage and necessitate eradication and quarantine efforts as soon as the responsible pathogen is identified.

8. Some plant diseases are expensive and slow to eradicate; success depends on having a strategy in place for eradication and acting on the strategy at the first detection of the disease.

9. Because of the long lag between introduction and detection of some plant diseases, eradication can be successful if the disease does not spread too far from the initial source by the time that the eradication effort is initiated. The epidemiology of the disease must be considered when determining whether or not to attempt eradication (Jeger et al. 1998; Madden and van den Bosch 2002). Considerably more experimental and theoretical research is needed on the pathogens posing a threat to prepare strategies for control, containment, or eradication.

10. Many plant diseases are of great economic importance because of international trade.

11. Most citizens of the United States are not directly involved in agriculture, and there can be an uneasy relationship between the agricultural industry and the rest of society. Homeowners, for example, might not be tolerant of the efforts needed to control or eradicate a crop disease. Educational efforts are needed to inform the public about the need to protect our crops if attacked by biological weapons. Lawsuits can considerably limit a state's ability to carry through on appropriate control or eradication actions. The political will to proceed with eradication in the face of these complications is needed.

12. Although identification of particular strains of pests and pathogens with molecular and other techniques is possible, such procedures may not be broadly available for others. Laboratories that can provide rapid molecular characterization of plant pathogens for many crops and pathogen types need to be identified and supported.

SUMMARY

The committee's in-depth study of various agents raises important issues relevant to countering agricultural bioterrorism. The committee has identified some of the scientific and technical capabilities and needs of the US system for defense against intentional biological threats to agricultural plants and animals. Through the presentation of lessons learned from its analysis, the committee hopes to have highlighted the importance of a solid, science-based defense system for US agriculture.

4

Research Needs and Opportunities

INTRODUCTION

There are needs and opportunities for aggressive research in both science and technology to improve our ability to prevent, detect, respond to, and recover from biological attacks on agricultural plants and animals. The scientific knowledge and the technological developments for protecting plants and animals against naturally occurring or accidentally introduced pests and pathogens constitute a starting point for these efforts—but only a starting point—and there is much more to be done (Chapter 5, Finding I.H). For example, our knowledge of even the most common agricultural pests and pathogens is still incomplete, to say nothing of the problem presented by new or rare organisms that might be introduced in a deliberate attack (Chapter 5, Recommendation I.F.1.a). Molecular techniques have not been applied to the study of many of them. Moreover, there are gaps in our knowledge of the epidemiology and pathogenesis of these organisms, both common and uncommon (Chapter 5, Recommendation I.F.1.b). The challenge is particularly difficult because of the large number of species (and cultivars or breeds) of plants and animals that make up the US agricultural industry, each one of which can host a unique and ever-evolving group of pests and pathogens. Research must now also consider the unique challenges presented by deliberate introductions, including those which use genetically engineered organisms.

With respect to the tools at our disposal for coping with a biological attack, although we must continue to make efforts to develop more-sensitive technologies for border interdiction of surreptitiously introduced organisms, there appear to be serious practical limits to the effectiveness of that approach (Chapter 2, Chapter 5, Finding I.D). That creates an even more urgent need for instruments and systems that will improve our ability to detect and identify an organism and diagnose the resulting disease or infestation in the field at the earliest possible time after its introduction (Chapter 5, Recommendation I.F.2.a). This will require more-sensitive detection and identification systems that are capable of increasingly rapid analysis. These systems will also have to be portable and simple enough to be directly deployable in the field, or integrated in a system that will facilitate analysis of any field sample anywhere in the country in a matter of hours (Chapter 5, Recommendation I.E.1.b). Beyond the issue of detection and identification, analysis in the event of a deliberate biological attack must serve a forensic function by identifying the source of the pest or pathogen—an even more challenging problem.

Finally, there is the issue of recovery from an attack. A great deal of effort has gone into developing procedures for limiting the spread and effects of plant or animal diseases and infestations, from quarantine to sterilization and cleanup. Among the lessons learned is that those procedures must be tailored to the specific organism or toxin involved. Work is now needed on the new biological threats to agriculture that we may face in an attack (Chapter 5, Recommendation I.F.2.c).

If the challenges are great, however, so are the opportunities. The extraordinary advances in molecular and cellular biology, in genomics, in sensor development, and in information technology (including the hardware and software for gathering, processing, and distributing information), all hold promise for improving our capacity to deal with biological threats. The key is establishing research and development priorities for the near term and the intermediate term. For example, in the absence of complete genome sequencing for all possible threat agents—an impractical short-term goal—what are the most important organisms to sequence, or what subset of those organisms' genes would be most useful? Are there approaches to establishing adequate, if incomplete, genetic identification that would facilitate early action to contain a disease or infestation? Are some genes more promising than others for use in identifying the source of an agent?

There is a similar need for establishing priorities for developing cellular and biochemical identification techniques, such as nucleic acid-based technologies or metabolic pathway analyses. Here, new analytic techniques are likely to be valuable, but it will be important to develop criteria for the sensitivity of such analytic tools for the organisms of greatest concern.

Each of those kinds of R&D can be helped enormously by information technology—the development of sensors, the storage and mining of data, and the various communication systems that are key to constructing an effective and rapid response plan (Chapter 5, Recommendation I.F.2.b). Therefore, there is also a need to factor in the new developments in information technology in setting the R&D priorities for dealing with biological threats to agriculture.

In addition to the molecular sciences, it is important to also recognize the role of the ecological sciences in helping to understand how to respond to and eliminate agricultural pests and pathogens. For example, birds or insects could play key roles in transmitting agents and causing widespread distribution. Understanding the ecology and behavior of vectors for plant and animal pests and pathogens is extremely important for developing countermeasures to agricultural bioterrorism.

The introduction of a pest or pathogen into agricultural or natural environments is an ecological process and can be broken down into four phases: dispersal, colonization, survivorship, and population growth. Because bioterrorism can be countered by disrupting any of these four, knowledge of each phase is needed.

Perhaps the greatest challenges lie in understanding human behavior. The social sciences, coupled with public health, play a crucial role in evaluating the human dimensions of agricultural bioterrorism. This includes assessing the costs to producers and consumers; developing effective approaches to communicating with producers, public officials, and the public at large; and identifying ways of assisting affected communities (Chapter 5, Recommendation I.F.1.c).

In this chapter, knowledge in each of those fields is reviewed and the potential for advances is discussed. Several R&D directions would enhance our ability to defend against agricultural bioterrorism, but resources are always limited, and not all avenues can be pursued. Therefore, R&D priority setting is a necessary part of an overall defense plan. This chapter can serve as a starting point for such an effort, as it recommends several key subjects for future R&D (see Research Needs later in this Chapter; Chapter 5, Recommendation I.F).

DETECTION, IDENTIFICATION, AND DIAGNOSTICS

Early detection, identification, and diagnosis are pivotal for limiting the extent of an outbreak, whether it is natural or intentional. Technology must be rapid, field-deployable, accurate, and sensitive; and it should be inexpensive and require little training for use. The limitations of individual detection methods should be recognized. For example, hand-held automated immunoassays—such as those used for environmental detection of anthrax—although rapid and field deployable, can lead to false positives and over-interpretation of results. Yet, they have a place in the overall scheme for detecting and identifying biological agents, as they signal the need for more rigorous and slower identification methods. It is clear that a variety of detection, identification, and diagnostic tools are needed to achieve early warning.

There have been important advances in technology for detection and diagnosis of microbial diseases, but many have not included agriculturally important pathogens; nor have they been inexpensive or field-deployable. The task is substantial. Classically, microbial detection and identification involved growth in culture and the use of differential metabolic assays to determine the species of most bacteria, or the use of cell culture and the electron microscope to diagnose viruses (and some bacteria). Most environmental samples have had to be cultured to obtain enough organisms for reliable identification. The time (4 hours to 30 days) and facilities (laboratories, equipment, and trained personnel) required for this approach can be prohibitive, especially in the face of a multisite outbreak.

Detection, identification, and diagnosis are based on several steps, including sample collection, recognition of the agent, disease, or infestation, and reporting the recognition. Sample collection can be part of a routine and continual program, but it usually depends on a signal that something is wrong. The signal can come from passive surveillance (for instance, a farmer notices a problem) or active surveillance (as in a formal monitoring program). In the case of crops, early detection is limited by the low surveillance of agricultural fields. With 60 million acres of wheat, for example, disease outbreaks are typically discovered only after several cycles of infection have occurred and disease incidence is too high to consider eradication. Improved methods of crop surveillance must be developed to discover diseased plants sooner after pathogen introduction (Chapter 5, Recommendation I.F.2.a). Improvements can include greater industry awareness to identify a potential pest or pathogen outbreak, the development of more-sensitive remote sensing technologies, and better use of modern survey sampling methods (Binns et al. 2000) to monitor fields for unusual signs. For both plant and animal pathogens, biosensors in farms or fields have the potential to serve as tools for initial warning and to indicate that more sample collection and analysis are needed. However, no field-tested and validated systems are commercially available for this purpose, and more research is needed for development of such systems.

Specific methods for detection and identification and then laboratory surveillance and diagnosis are discussed below.

Detection and Identification

Recognition technologies include those based on nucleic acids, antibodies and antigens, ligands and receptors, and physical properties (reviewed in NRC 1999). Nucleic acid-based technologies rely on the extreme selectivity of DNA and RNA recognition and sensitivity enhanced by selective amplification. Binding can be detected easily, for example by tagging with a reporter molecule that fluoresces. The main advantages of this technology are that all living organisms have DNA or RNA (some portions of which are unique to the organism), very small amounts can be detected, a sample can be probed for many different sequences simultaneously, and several commercial products are available.

An added advantage of nucleic acid-based systems is that some portions of the sequence may change with time and host, offering a forensic trail. Efforts are under way to enhance our capabilities for using molecular information (such as DNA sequence) to trace back and determine the origin of biological threat agents (Murch 2001b), and high-throughput laboratories for such capabilities have been proposed (Layne and Beugelsdijk 1999). However, as illustrated by our nation's troubled attempts to determine the source of the anthrax used in the fall 2001 attacks, the utility of molecular forensics to identify perpetrators, although useful, is modest (Weiss 2002, Enserink 2002a).

Disadvantages of nucleic acid-based technologies include the difficulty of isolating the DNA or RNA in environmental samples, degradation of the nucleic acids in the environment, and interference by related sequences or contaminants. The sensitivity in detecting a microbial pathogen is 10^3 – 10^5 colony-forming units,[1] and the time required depends largely on that required for sample preparation, which can vary from hours to days. Hand-held polymerase chain reaction (PCR) units and DNA and RNA chips (Marshall and Hodgson 1998) are available commercially. One PCR-based system has been developed for detecting and identifying foot-and-mouth disease (FMD) virus in ambient aerosols and animal body fluids (Northrup 2001). The system is being tested and evaluated for field deployment.

Other nucleic acid-based technologies, not based on PCR, are in early stages of development. For example, a recent report illustrates a technique that uses the following scheme: single-stranded DNA is captured on a glass surface and then probed with complementary strands of DNA attached to gold nanoparticles, and finally the gold particles are used to carry an electric current (Park et al. 2002). The resulting electronic chip has the potential to detect DNA from organisms, such as anthrax, in minutes and appears to be more sensitive than other high-speed techniques; however, much more testing and validation are needed before this novel technique is field-deployable.

Antibody-based technologies recognize specific sites or components on the microbial cell, and antibodies can be made for any microorganism if it can be obtained in pure culture. Because of the expense and difficulty of making monoclonal antibodies, it is useful to provide methods for disassociating the antibody from the microorganism for reuse. The binding of the target (antigen) to the antibody can be monitored with several reporting methods. Alternatively, a second antibody labeled with a fluorescent dye can be bound to the antibody-target complex or to another site on the target in a sandwich assay.

Disadvantages include nonspecific binding, cross-reactivity among closely related organisms, degradation of antibodies over time, poor reproducibility of antibodies, and difficulty of producing the target antigen in pure culture. In addition, some target sites may be lost after the microorganism has been propagated over several generations. Nevertheless, some of the most sensitive technology is based on antibodies.

Fluorescence-based immunosensors have a sensitivity of 10^4 microorganisms/mL, and immunoelectrochemical sensors and sandwich assays have a sensitivity of 10^3 microorganisms/mL. Increased sensitivity is gained with immunoPCR,

---

[1]Sensitivities are based on the committee's experience in detecting biological agents in samples collected from the environment. These values might be lower for laboratory-prepared samples.

which tags the antibody with a short strand of DNA that takes advantage of amplification of the antibody-target complex (Joerger et al. 1995).

Ligand-based technologies solve many of the problems of antibody-based technologies and enhance the specificity. They are based on the principles that there are cell-surface proteins peculiar to a particular cell type (the same assumption as in the development of antibodies) and that ligands that specifically bind these proteins can be found. Ligands may be specific to serotypes or common to related groups. Dyes that are structural analogues for receptors have been used historically in classical microbiological assays. More recently, with the development of combinatorial chemistry, ligands directed at specific receptors have been used to specifically bind a number of pathogens, either as vegetative cells or as spores (Powers and Ellis 1998). Exotoxins and viruses have also been bound by ligands; this is important for genetically engineered organisms, in as much as these ligands recognize the toxin or other pathogenicity marker regardless of the organism used for production and delivery. Difficulties include interference and competition with natural ligands and the fact that receptor sites under gene regulation may alter their expression state under various environmental conditions. Sensitivities as low as 10^2 microorganisms/mL have been reported in environmental samples.

Reporter technologies are based primarily on the generation of one or more of three kinds of signals: electrochemical, piezoelectric, and optical. Electrochemical transduction uses biochemical reactions to generate an electrochemical signal. Measurements of 10^3 microorganisms/mL have been demonstrated in 1-3 minutes with antibody recognition.

Piezoelectric transduction uses special crystal materials that produce an electric charge when pressure is applied. The application of an electric current causes them to vibrate at a frequency that depends on their dimensions (mass). For example, binding antibodies or ligands at the surface of the crystal alters the mass and hence the vibration frequency. The binding of the target to the antibody or ligand surface coating causes a further change in vibration frequency. The sensitivity is 10^4–10^5 microorganisms/sample. Piezoelectric transduction is the basis of the chemical agent detectors that use surface acoustic wave (SAW) technology (Guilbault et al. 1992). Optical transduction includes light absorption (colorimetry), light emission (luminescence), and light scattering. Luminescence, especially fluorescence, is sensitive enough to detect a single molecule. One energy of light is used to excite the fluorescent molecule, and relaxation to its original state involves the emission of light at lower energy. Those energies depend on the chemical composition of the substance, so it is possible to excite and detect specific substances in cells. There are also problems with overlap of fluorescence signals in biological samples and with background fluorescence and other interferences. Nevertheless, fluorescence offers the most sensitive of the reporter technologies and has been widely used in environmental and medical applications.

Automated fiber-optic biosensors that require little sample preparation and can detect various biological agents rapidly and simultaneously are being field-tested and deployed (Ligler et al. 1998, Ligler 2000)

A few technologies that do not fall into one of the above categories merit consideration. Those based on such physical properties as mass, charge, and size are examples. In mass spectrometry (MS), a substance is converted to gaseous ions, and the fragments are separated on the basis of their mass-to-charge ratio. The spectrum of fragments is characteristic of specific components, such as biological molecules, and can be used for identification. An MS-based system that integrates sample collection and detection capabilities has been developed and is being tested for use on biomarkers indicative of biological terrorism agents (McLoughlin et al. 1999). Quantities as small as 10^5 molecules and 10^6 microorganisms have been detected. However, heterogeneous samples require separation methods (such as gas chromatography) before they can be reliably identified with this technology.

Other separation methods have also developed for microbial detection, such as sedimentation followed by ultrafiltration. The final stages of this method involve electrospray aerosolation of the filtrate, differential mobility analysis, and condensation nucleus counting. The sensitivity is reported as 10^3 particles within an hour.

Detection and identification of pathogens in the environment pose difficulties for any technology (Walt and Franz 2000). The sensitivity or detection limit needed depends on how much of the pathogen is required for an infectious dose (for example, some human pathogens have an infectious dose at which 50% of the population becomes infected with 100 microorganisms), the conditions under which the determination must be made, and the time in which the determination must be made. The Department of Defense has some capability using newer emerging technology (Walt and Franz 2000). Several promising technologies and methods are under development (Dutton 2002). For highly infectious agents, sensitivity has been a problem (that is, it might not be possible to detect a dose that has important health consequences). What is particularly important is that few rapid and field-deployable detection and identification methods are available for plant and animal pathogens or pests (Chapter 5, Recommendation I.F.2.a).

Laboratory Systems and Diagnosis

Several sample collection and analysis steps need to be coordinated to detect and identify biological agents in the environment or in their hosts accurately and reliably. Laboratories need to work together to track the prevalence and spread of biological agents on geographic and temporal scales. Likewise, communication between laboratories is vital for rapid response measures. There is no coordinated national system for laboratory analysis of biological threats directed at US agriculture. In the event of a large-scale or multi-focal attack, or one using agents that persist in the environment or spread rapidly, existing laboratory systems and resources would be insufficient for minimizing the spread and impact. In light of the above, the United States needs a laboratory system that can rapidly detect, identify, diagnose, and respond to agricultural bioterrorism threats (Chapter 5, Recommendation I.E.1.b). One possible model system that could be adapted for this purpose is the public-health laboratory response network (LRN) that has been established to respond to human health threats (CDC 2001b). This network is described briefly below; the committee recommends an analogous system for agricultural bioterrorism in Chapter 5.

The LRN is a consortium of public health laboratories (clinical, local, state, federal, and specialty) organized to provide state-of-the-art diagnostic and environmental testing support for bioterrorist threats to humans throughout the United States. The network is supported under Centers for Disease Control and Prevention (CDC) cooperative state and local bioterrorism preparedness and response agreements and is managed through a contract with the Association of Public Health Laboratories. It can operate on clinical, environmental, and forensic samples. The LRN can respond to large numbers of clinical samples by sharing the workload. An almost unlimited number of environmental samples can be handled by distributing them to a combination of laboratories as described below. Testing in support of law enforcement includes training in and execution of chain-of-custody procedures.

The concept of operations of the LRN is a tiered system consisting of four "levels". Level A is the designation for clinical and small local public health laboratories that are able to perform basic bacterial isolation and morphological studies to "rule out" the presence of an important agent. Level B is the "confirmatory" level at which larger local and state public health laboratories are able to perform definitive identification of threat agents by using reagents developed and distributed by the LRN. A clear example of the functioning of this combination was illustrated by the diagnosis of the first case of inhalation anthrax in Florida, where the level A hospital laboratory forwarded a suspect bacterial culture to the level B state laboratory for identification. The agricultural diagnostic laboratories could be structured in a similar relationship.

Level C consists of larger state public health laboratories that perform complicated diagnostic procedures, such as botulism testing that requires the use of mice. Level C laboratories have the ability to characterize, classify, and compare bacterial isolates by using molecular markers and to perform rapid screening tests with advanced technologies. Sharing such data across the LRN can be the basis of recognition of an outbreak as opposed to a single isolated case of a disease. CDC, the US Army Medical Research Institute of Infectious Disease (USAMRIID), and the Naval Medical Research Center (NMRC) all have level C capability on a 7 days per week, 24 hours per day basis as needed.

Level D laboratories consist of the specialty units in CDC, USAMRIID, NMRC and the national laboratories that perform advanced forensic testing on specific threat agents. These laboratories also have the lead in development and transfer of advanced diagnostic methods to level B and level C laboratories. Comparison of the strains from the anthrax cases of fall 2001 was done at CDC, USAMRIID, The Institute for Genomic Research (TIGR), and Northwest Arizona State University. The agricultural reference laboratories could be integrated into the LRN in a similar fashion.

Overall, the committee suggests that a similar network be established for agricultural pests and pathogens (Chapter 5, Recommendation I.E.1.b). This network should be capable of integrating diagnosis and response information (for example, protocols for response actions), research information (for example, genomic sequences), and survey information (for example, recognition of strange pest or pathogen occurrences). It is possible that subnetworks will need to be designed for the various components of the network in order to efficiently manage its multiple roles. The CDC's LRN costs approximately $20 million per year to operate and maintain, with an additional cost of $8 million per year for the local labs involved in the network. It is not clear whether the costs would be similar for an agricultural system. The LRN currently discloses the agents for which it tests. Dialogue about the advantages and disadvantages of keeping this information confidential for an agricultural bioterrorism network should take place.

Although the LRN proved effective in confirming suspicious samples during the anthrax attacks in fall 2001, for consequence management, greater laboratory capacity was needed. Numerous environmental samples are involved in testing for spread of biological agents and in determining the effectiveness of cleanup. It is important that such laboratory networks also be supported by other laboratories that can assist in such situations. The development of surge capacity should accompany the development of an agricultural laboratory network (Chapter 5, Recommendation I.E.1.b).

GENOMICS, GENOME SEQUENCING, AND TRANSGENIC TECHNOLOGIES

Probably no field of research in the life sciences opens the door to more strategies and offers more tools for countering agricultural terrorism than genomics, genome sequencing, and transgenic technologies (Gibson and Muse 2002). With entire sequences and high-resolution physical maps now available for the genomes of humans and several model organisms—including the mouse, the fruit fly, the nematode *Caenorhabditis elegans*, yeast, and the seed plant *Arabidopsis thaliana*—science and technology have turned increasingly to sequencing the genomes and understanding the functions of genes of economically important microorganisms, arthropods, plants, and animals. This information is becoming the foundation for understanding pathogenesis in the case of pathogenic microorganisms, feeding ability and preferences in the case of insect pests, development of new genetic mechanisms of resistance or innate immunity in the cases of agriculturally important plants and animals, and development of new vaccines in the case of animal pathogens (Chapter 5, Recommendation I.F.1.a and b).

Genomic information also forms the cornerstone for developing detection and identification methods (see Detection, Identification, and Diagnostics in this Chapter). Comparative genomics, enabled by having the sequences of several pathogens or pests, is not only critical for forensics, but for diagnosis and identification. One of the major constraints to detection, identification, and diagnostics is that there is not enough information to efficiently design tools that accurately and rapidly detect and identify many potentially-weaponized agricultural pests and pathogens. Sufficient information is needed to select appropriate DNA sequences for primers, or proteins as targets for antibodies, so that we can make better use of those important tools.

Mammal Pathogen Genomics

Much of the practical application of gene and genome sequences to potential agents of bioterrorism has already occurred in the field of human pathogens. Several pathogens have been sequenced, and primers and probes are available for real-time PCR diagnostic tests. Since the anthrax outbreaks in October 2001, work on sequencing the genomes of *Bacillus anthracis* (Enserink 2002b) and agents of several other potential bioterrrorism diseases—such as those causing brucellosis, glanders, Q fever, and plague—has intensified (NIAID 2001, TIGR 2002). In addition, databases have been established to compare the sequences and functions of smallpox genes with those of other poxviruses (NIAID 2001).

Several foodborne pathogens carried by poultry and livestock, such as *E. coli* 0157:H7 and *Campylobacter jejuni,* and the important animal pathogen *Pasteurella multocida* have been fully sequenced (TIGR 2002, May et al 2001).

Molecular characterization based on gene sequences has been used to identify and characterize several animal pathogens, including the Nipah virus (Enserink 1999). In that case, full viral genome sequencing was used 1) to show the relationship of the new virus to the previously characterized Hendra virus and 2) the agent's spread through human and animal hosts. Continued work of this nature on agricultural threat agents will be needed to bring diagnostic and forensic tests for agriculturally important pests and pathogens into the 21st century. Primers and probes for real-time PCR, especially those designed for pathogenicity genes, are not available for most agricultural pests and pathogens (Chapter 5, Recommendation I.F.2.a). A database of gene sequences that are indicative of biological threats to agriculture would be useful for rapid analysis of samples and design of diagnostic tools before and during an outbreak (Chapter 5, Recommendation I.F.2.b).

Plant Pathogen and Pest Genomics

At this writing, sequences of the entire genomes of only two nonviral plant pathogens are available, namely *Xylella fastidiosa* (Simpson et al. 2000), a bacterial pathogen of citrus, and *Agrobacterium tumefaciens* (Goodner et al. 2001; Wood et al. 2001), the cause of crown gall on a wide variety of dicotyledonous plants and the vector used in laboratories throughout the world to introduce transgenes into plants. The entire genome of no fungal plant pathogen, plant parasitic nematode, or arthropod pest has been sequenced, although considerable information is available or emerging on the sequences and functions of key genes used by these pests to attack plant hosts and on key genes used by plants in defense against these pests (Keen et al. 2000). Partial or entire sequences of the genomes of several plant viruses are available.

Recent research on the plant pathogenic fungus *Fusarium graminearum* (O'Donnell et al. 2000) illustrates the kind of information that can be obtained by using molecular phylogenics as a source of information on the possible origin of a new disease outbreak, to help to ensure that control measures are appropriate for a strain responsible for the disease. *F. graminearum* causes a disease of wheat heads and kernels known as head blight or scab and is one of the most destructive pathogens of wheat worldwide. In addition to poor yield and shriveled and discolored grain, the fungus produces mycotoxins that make the grain unsuitable for feed or food. Starting with 37 single-spored cultures of the fungus obtained from wheat and other host plants worldwide and working with six single-copy genes from strains selected to represent the global diversity of the species, O'Donnell et al. (2000) were able to group the cultures into seven biogeographically structured phylogenic lineages (clades). That all six molecular genealogies demonstrated the same seven biogeographically structured lineages suggests that gene flow between or among lineages has been limited during their evolutionary histories. Pan-Northern Hemisphere, Asian, South-Central American, South American, and three African clades were represented. With that information, every new occurrence of *Fusarium* head blight can now be characterized on the basis of the biogeographic lineage of the pathogen, and this helps to clarify the source of the pathogen. Although the work was initially done for unintentional outbreaks, such information can be critical when attribution is an important component of effective response to and recovery from an intentional biological attack (Chapter 5, Recommendation I.F.2.a and b).

Breeding and Transgenic Technology

Plant breeding and animal breeding have gone on since the beginning of agriculture, but with increasing precision and sophistication starting with the work of Gregor Mendel in the late 19th century, and now with the aid of modern biotechnology at the beginning of the 21st century. Knowledge of the function of key genes and possession of high-resolution physical maps of the locations of these genes lay the foundation for new and more-precise approaches to development of pest- and pathogen-resistant crops and farm animals, new biological control agents, and new vaccines.

How the new information and technology might be used to counter a bioterrorist threat agent in agriculture will depend on the threat agent and the targeted crop plant or animal. In the case of crop plants, for example, a genetic marker that maps close to a plant gene for pest or pathogen resistance can be used to select for resistant genotypes of the crop plant without having to screen segregating populations of plants with the agent itself. That can permit breeding for pathogen or pest resistance (for example, for soybean rust) as a means of deterring or nullifying an introduced threat agent. Recent evidence shows that genes used by plants for recognition of and response to their pests and pathogens—the so-called R genes—make up a family of similar and highly conserved genes used for recognition of viruses, bacteria, fungi, nematodes, and insects (Keen et al. 2000). That raises the prospect of development of crop plants with broad resistance to both existing and new sources of virulence that might develop or be introduced into a population of plant pests or pathogens (De Wit et al. 1998). R genes have been engineered into tomato and overexpressed to lead to broad-spectrum disease resistance (Oldroyd and Staskawicz 1998). Likewise, overexpression of a regulatory gene involved in systemic acquired resistance confers broad-spectrum disease resistance in *Arabidopsis* (Cao et al. 1998). Such research, included in comprehensive efforts to understand the genetics of host-pest interactions, needs support as part of a long-term approach to the deterrence of, response to, and recovery from attacks on US crops.

Transgenic technologies, also known as genetic engineering and genetic transformation, constitute an application of the knowledge of genomes and gene function with the potential to reduce or eliminate the risk posed by threat agents. The expression in plants of a gene from the insect pathogen *Bacillus thuringiensis* (*Bt*) for production of one of the endotoxins used by *Bt* to kill endotoxin-sensitive insects has proved highly effective as a source of resistance to the European corn borer and several insect pests of cotton (NRC 2000). Different *Bt* genes and their gene products are available or could be made available for targeting insect pests that might be used as threat agents against US crops. Many kinds of genes involved in disease resistance and defense responses are being tested as transgenes to protect crop plants against fungal (and other) pathogens (Broglie et al. 1991; Lorito et al. 1998). Thousands of field trials of transgenic crops are taking place. Although there is an initial cost in upfront research and development, pest-protected transgenic crops have been marketed and used on wide geographic scales in the United States, presenting a feasible, and in many cases, practical, alternative to chemical control (NRC 2000).

One of the most promising uses of transgenic technologies is in the control of plant viruses via "coat-protein-mediated resistance" (Beachy et al. 1990). The transgene is a gene from the target virus itself, specifically, the gene used by the virus to produce the coat protein that is wrapped around its genome and is responsible for its crystalline appearance. The production of the same coat protein by the plant provides immunity to the virus in direct proportion to the amount of coat protein produced. Most applications of this technology have been against insect-vector plant viruses, including several viruses of cucurbits, potato leaf roll, and papaya ringspot virus (Gonsalves 1998). Indeed, this application of transgenic technologies has the potential to control many of the agriculturally important insect-vector plant viruses for which the only controls available now are pesticides targeted at the vector.

The safety of transgenic crops to the environment or humans continues to be questioned, but studies conducted by the National Academy of Sciences (NAS 1987), the National Research Council (NRC 1989, 2000), and the Organization for Economic Cooperation and Development (OECD 1993) have consistently found that transgenic crops raise no new categories of safety concerns beyond what might be the case for crop plants genetically modified through conventional breeding. Moreover, with all applications of genomics, gene sequencing, and transgenic technologies, research is needed not only to counter threats of agricultural bioterrorism but also to overcome naturally occurring plant and animal pests and pathogens as a contribution to sustainable growth in agriculture—a win-win investment (Chapter 5, Recommendation I.F.2.d).

INFORMATION TECHNOLOGY

The ultimate tool for agricultural pest and pathogen management would combine everything we know from all sources to prescribe the best course of action and would stay abreast of developments. It would allow an individual specialist or even a farmer to describe symptoms observed in a corner of a single field so that a potential pest or pathogen outbreak and an appropriate course of action could be identified. It would 1) combine the field operations with thousands of others to synthesize a regional, national, or global picture of a given problem at any moment; 2) correlate many such problems with each other and with other data—such as rainfall, temperature, and chemical use—to facilitate improved resource management and response to problems; and 3) integrate physical and biochemical information on samples of the pest or pathogen collected at each site in an attempt to understand natural spread and evolution and to identify modes and routes of dissemination of organisms used for agricultural warfare or terrorism. Is such a tool feasible today?

The discipline of biological informatics (bioinformatics) has emerged in response to recent data explosion in the biological sciences, and it is rapidly becoming a central player in "the new biology." *Informatics* has many definitions; all encompass aspects of data acquisition, manipulation, analysis, and presentation. Although traditional approaches focus on the information content, informatics focuses on the structure of the information, not the content. Study of the structure of information leads to the invention and implementation of algorithms, and these lead to efficient and flexible methods for extracting desired information, presenting it when and where it is needed, and using it to facilitate intelligent decision-making.

It is convenient to think of informatics as a system that extracts useful information from massive and frequently unrelated data sources. Decision support systems are the other critical part of an overall strategy for making use of data. Decision support systems are often described as user-friendly front ends of otherwise unintelligible data systems. As tools that allow a non-specialist, non-programmer to answer practical questions on the basis of available data, decision support systems often use complex models and statistical approaches, expert systems, knowledge bases, and databases in the presentation of questions and results.

Originally the domain of academics and researchers, the World Wide Web has become the method of choice for almost all information sharing. Browser-based interfaces for examining data and using decision support systems reduce software development costs, software maintenance complexity, and the computer hardware required for many applications. The web also allows information from many sources and locations to be integrated and presented seamlessly to the end user.

Creating an all-encompassing Web-based system for agricultural management is a formidable task. A similar situation exists for human health management, and efforts are being made to develop such a system (see Detection, Identification, and Diagnostics in this Chapter). However, many resources for agriculture are available on the Web. For example, basic biological data are available on many plant and animal pathogens. GenBank houses the National Institute of Health collection of publicly available DNA sequences; it is maintained by the National Center for Biotechnology Information (NCBI) in Bethseda, MD. GenBank contains over 14 billion bases from approximately 100,000 species and is doubling in size every 14 months. A typical GenBank entry contains information on a gene locus and definition, a unique accession number, organism information, literature citations, revision notes, such biological features as coding regions and their protein translations, and, of course, the DNA sequence itself. GenBank and related general resources—such as the DNA Databank of Japan (DDBJ), at the Japanese National Institute for Genetics, in Shizuoka, Japan, and the European Molecular Biology Laboratory databank, maintained by the European Bioinformatics Institute, in Cambridgeshire, UK—form the International Nucleotide Sequence Collaboration. In addition, several sites, such as the US Department of Agriculture Agricultural Research Service genome database, maintain genome information specific to agriculturally relevant plants and their pathogens. Several data centers tailored for specific types of pathogens are also available. One center deals with host-pathogen relationships for plant-parasitic nematodes and includes plant susceptibility, related bacterial and fungal interactions, and geographic information. The Ecological Database of the World's Insect Pathogens examines fungi, viruses, protozoa, and bacteria that infect a variety of arthropods. Plant Viruses Online documents plant viruses and their effects on many plant species. The Agriculture Network Information Center has a useful list of links to agricultural databases. Similarly, some sites attempt to integrate information on relevant physical characteristics—such as rainfall, temperature, and sunlight—and use Geographical Information Systems (GIS).

Given the apparent wealth of information, two features are still required to create a universal system: the disparate data sources need to be combined with each other, and up-to-date tracking of pests and pathogens of interest must be integrated with these sources. Informatics is key to bioforensic capability. Although limited or local in scope, first steps have been taken toward a system that integrates information on plants, their pests and pathogens, and the weather. The Integrated Pest Management system (IPM) at Oregon State University (Coop 2000) is an example.

Steps are needed to develop a coordinated information technology base for agricultural bioterrorism (Chapter 5, Recommendations I.E.1.d and I.F.2.b) that would include 1) information on gemonic sequences for bioforensics and research, 2) detection and diagnostics obtained by laboratory networks, 3) epidemiological analyses, 4) outbreak-control information, 5) a directory of experts in related subject matter, and 6) relevant data from R&D projects, field stations, regional agriculture laboratories, and agriculture professionals.

CONTROL TECHNOLOGY

Once a pest or pathogen has been introduced into US agriculture, some form of control technology must be applied (Chapter 5, Recommendation I.F.2.c and d). The specific technology used will depend on the agent, including our ability to correctly identify it. Tactics for controlling pests and pathogens of agricultural crops and animals include two basic options: eradication and management.

Eradication

Eradicating a pest or pathogen is eliminating it from the environment (at least until the next introduction). Eradication can be difficult and expensive, especially if the pest or pathogen is soilborne or has the ability to spread rapidly and widely. Some pests or pathogens may not be amenable to eradication, because we lack the necessary tools and information.

Tools that can be used in eradication programs include the following:

- Correct and rapid identification of the threat agent in the host or environment (see Detection, Identification, and Diagnostics in this Chapter), so that it can be known with confidence whether the agent is still present. For example, the fall 2001 program to eradicate anthrax spores from the Hart Senate Office Building required multiple applications of fumigant because of repeated detection of the agent.

• Efficient survey and mapping methods to detect the extent of the infection or infestation so that planning and execution of an eradication program is effective.

• In the case of infected plants or crop damage caused by pests or pathogens, ability through sensing, including remote sensing, to make it possible to pinpoint areas of infection or pest infestations in a defined geographic area rapidly and accurately. Remote sensing has been used to identify virus-infected fruit trees before the appearance of symptoms but has been less effective or even ineffective in distinguishing areas of pest damage from drought or other stresses in crops. New sensing technologies are needed that can recognize specific pests or pathogens on plants, ideally through aerial or even satellite imagery.

• Quarantine prevents the unintentional spread of the pest or pathogen to areas outside the infested or infected area while eradication procedures are being conducted.

• Removal and killing of diseased or infested animals or plants. This requires that the target agent be correctly and rapidly identified and that adequate resources be available for that purpose before the agent is so widely distributed that it is no longer logistically or economically feasible to carry out the eradication program.

• Administration of effective vaccines to reduce or eliminate the disease in farm animals. This implies that vaccines can be obtained and deployed in a timely manner. Having viable and adequate stockpiles of key vaccines might be crucial to a timely response, especially if eradication is the goal (Chapter 5, Recommendation I.F.2.c).

• Appropriate epidemiological models to ensure that a vaccination program, cull and slaughter, or a roguing program (a program to remove and destroy infected plants) will be effective (Chan and Jeger 1994; Ferguson et al. 2001; Madden et al. 2000; Madden and van den Bosch 2002; van den Bosch et al. 1999). Critical questions include these: How large should a buffer zone be? How rapidly must vaccination, cull and slaughter, or roguing be carried out? How far and how quickly can a pathogen or pest spread? Are there asymptomatic carriers of the pest or pathogen? What is the host range? How is the pest or pathogen transmitted? Are insect or other vectors important in transmission? If so, how is the pest or pathogen maintained and transmitted in the vectors? Models will need to be based on the fundamental population biology of the pest or pathogen (Sakai et al. 2001). Models were used in the UK FMD outbreak to evaluate risk-mitigation strategies before implementation; the models accurately predicted the eventual spread of FMD on the basis of the culling and vaccination strategy that was used (King 2002). Given the potential for this approach, more research is needed to develop and test epidemiological models for plant and animal pests and pathogens so that optimal eradication and containment strategies can be developed before a threat agent is introduced (Chapter 5, Recommendation I.F.2.d).

• Specific antibiotics, other chemotherapeutics, or pesticides, including fumigants. These products can be effective eradication tools, depending on the target (Dhadialla et al. 1998), and they are or can be legally authorized for use in the United States or subject to emergency registration if necessary.

• Sterile insect techniques, which have been used for agricultural pests and can be very effective.

• Eradication action plans that have been tested, ideally in the area of origin of the target pest or pathogen. Eradication plans, which are usually complex, as well as agent and host-specific, can be used to guide overall procedures (Chapter 5, Recommendation I.E.2.b). Successful pest or pathogen eradication efforts typically require detailed plans for quarantine, survey, and detection; specific eradication efforts; and educational programs directed at farmers, extension agents, and the general public. Specialized production facilities may be required to produce sterile insects or other tools for eradication. Such complex programs are unlikely to be initiated and accomplished in a timely fashion if plans are made on an ad hoc basis.

Managing Pests and Pathogens

Eradication is not now practical or feasible for many of the pests and pathogens (insects, mites or ticks, nematodes, weeds, and plant or animal pathogens) that might be used in agricultural bioterrorism. As a result, some pathogens or pests, if released, will probably become permanently established in the environment and in US agriculture. Such pests or pathogens could impose a serious economic burden on US agriculture, affect our competitiveness in world markets, and disrupt the social and economic fabric of rural communities.

Management tools for plant or animal pathogens or pests include a range of options:

• Pesticides or antibiotics or other chemotherapeutics, which can be applied to suppress pests or pathogens, although the emergence of resistance to these chemicals can be expected if adequate "resistance-management" tactics are not used (Casida and Quistad 1998). Timing the application of a pesticide or antibiotic requires effective methods for identifying the target agent and for monitoring the agent in a timely and cost-effective manner. The use of pesticides or antibiotics can result in concerns about nontarget effects, including the disruption of natural enemies by pesticides and the development of antibiotic resistance in nontarget animal or human pathogens.

• Vaccines, which might constitute the best method for management of animal diseases that cannot be eradicated or be part of an eradication strategy.

Vaccination programs are expensive and time-consuming. For some diseases, such as FMD, a ban on those animals or animal products by importing countries can be triggered if a vaccine is used for management. Some vaccines for important agricultural threat agents are stockpiled, albeit in relatively small quantities. More of these traditional (modified live or killed) vaccines could be produced and stockpiled immediately if necessary. Vaccines for some agents will be more useful than others in an outbreak. Finally, present technology generally does not support timely vaccine development and production after an outbreak is discovered (Chapter 5, Recommendation I.F.2.c).

In the short term, there are needs to identify threat-agent outbreaks that can be affected by a vaccination program and to analyze the cost effectiveness of developing additional vaccine stockpiles for those agents (Chapter 5, Recommendation I.F.2.c). Working with international trading partners on these projects may be feasible and more efficient. Exploration and development of newer vaccine technologies (such as DNA, vectored and replicon systems) should to be pursued; these technologies could be conditionally or fully licensed as generic systems now and then rapidly (in weeks) pushed to production as needed. Government support of basic immunology and vaccine research and development of the new technologies will have broad dual-use application in agriculture and pet animal health whether or not we ever need vaccines to control an intentional outbreak of disease in our livestock populations (Chapter 5, Recommendation I.F.1.a and b).

- Host plant resistance, which, like vaccination of animals, represents one of the best strategies for management of plant pests and pathogens. However, development of resistant commercial plant varieties adapted to the US environment takes time and may not be feasible in the short term, especially if traditional breeding practices are used. The use of molecular methods in developing resistant plant varieties can shorten the time required to obtain the desired resistance, and the use of transgenes can greatly broaden the range of pests and pathogens that can be managed with host plant resistance (see Genomics, Gemome Sequencing, and Transgenic Technologies in this Chapter). However, permission to release trangenic crops into the environment might require years of regulatory evaluation to ensure safety. Regardless, private and public investment in research or genome sequencing, gene function (functional genomics), and detailed characterization of gene products (proteomics) for key crop plants can be expected to accelerate and expand the ability of plant breeders to develop crops with resistance to pests or pathogens on the basis of modification of expression of plants' own genes or transgenes (see Genomics, Gemome Sequencing, and Transgenic Technologies in this Chapter; Chapter 5, Recommendations I.F.1.a and I.F.2.d).
- Quarantine, which can be used to contain some pests or pathogens in a specific geographic area. Quarantine requires monitoring and surveillance; efficient detection, identification and diagnostic methods; education of all personnel in appropriate management tactics; and legal authority to carry out the quaran-

tine. Containment, even if it only delays the spread of a pest or pathogen, may be cost-effective on a national scale, but cost-benefit analyses will be required to confirm the effectiveness.

- Classical biological control, whereby natural enemies from the geographic "home" of the pest are introduced. Some insect pests or weeds can be suppressed by the introduction of their natural enemies (parasitoids, predators, pathogens, and antagonists) (Bottrell 1998, Landis et al. 2000), and classical biological control has been directed against invasive insects, mites, and weeds for over 100 years. Other biological control approaches use, for example, nonpathogenic strains of microorganisms to outcompete or replace pathogenic strains on the host (Larkin and Fravel 1999). Classical biological control can result in the long-term suppression of the pest or pathogen population below the level where it causes economic injury. Classical biological control can be environmentally safe if adequate consideration is given to the specificity of the natural enemy introduced and if sufficient evaluation is carried out in a licensed quarantine facility by trained specialists. It can take years to complete the sequence of steps involved in classical biological control, so this approach is not an instant panacea, and not all potential bioterrorist agents could be suppressed by natural enemies released in such a program.
- Cultural controls, which are commonly used to manage serious pests or pathogens and which include timing of planting or harvesting, rotating crops, locating seed production areas away from an endemically infested area, providing refuges for natural enemies of the pests, coating seeds with a protectant before planting, producing pathogen-free seed through tissue culture, limiting access to the crop or farm animals so that pests and pathogens are excluded, and rearing "trap crops" that are more susceptible to the pest so that the pest can be destroyed before it invades the main crop. Cultural controls require a thorough knowledge of the ecology of the pest, the cropping system, and the willingness of farmers to modify production practices, so effective educational efforts are essential.
- Integrated pest management (IPM), a more holistic approach to pest management. IPM relies on the use of many tactics as appropriate and compatible to *suppress* the target pest or the damage it causes to a level that allows agricultural production to be maintained economically (Jacobsen 1997, Kogan 1998). IPM programs are information-intensive and require substantial efforts for deployment. An effective IPM program generally is crop-, location-, and pest-specific and often requires years to develop, validate, and deploy. If new pests become established in the United States as the result of a bioterrorism event, IPM programs that had been effective could be disrupted or made obsolete. Research necessary to develop and integrate the information and appropriate IPM management tactics for the new pest would be expensive and time-consuming, adding a burden to the research and extension community as well as to the agricultural production system.
- Epidemiological models, which can be linked to field data and used to

help to determine and test the optimal management strategies for diseases or infestations of plants and animals (Anderson and May 1991, Campbell and Madden 1990). Research on the further development of the models for threat agents is urgently needed so that plans can be put into place both for containment of introduced organisms and the integrated management of diseases and infestations that cannot be eradicated (e.g., Chan and Jeger 1994; Madden and van den Bosch 2002) (Chapter 5, Recommendation I.F.1.a). Future model development must focus on heterogeneity of host and environment, simultaneous development of disease or infestation in space and time, and metapopulation dynamics (to deal with discrete foci of infections).

Overall recommendations for enhancing the science and technology capabilities for control and management appear in the Research Needs section. New approaches and technologies, such as behavioral manipulation of insect pests or their enemies (Foster and Harris 1997), might be needed to stay ahead of the battle against damaging agricultural pests and pathogens, especially if these agents are designed by terrorists to overcome current control methods (Chapter 5, Recommendation I.F.2.d).

Disposal and Decontamination

Disposal of contaminated plant or animal material and decontamination of products, facilities, equipment, and in some cases, soil are critical components of response to and recovery from some agents. The disposal or decontamination procedures used and their effectiveness and acceptability are highly case-specific: they depend on the nature of the agent, the commodity affected, and the extent of disease or infestation. For example, FMD is so highly contagious that large numbers of infected and potentially exposed animals could be slaughtered and disposed of at the farm of origin. Mass burial and burning are the main alternatives for such disposal. Both are expensive, repugnant to many, and raise environmental concerns. The public made it clear, during recovery from the 2001 outbreak of FMD in the UK, that large piles of burning carcasses were unacceptable (King 2002). Novel methods for 1) carcass disposal and 2) inactivation of FMD virus in and on carcasses, or 3) alternatives to mass slaughter in FMD outbreaks are urgently needed. Decontamination of products, equipment, or facilities is less of a problem because FMD virus is inactivated by heat, irradiation, or chemical treatment at high or low pH.

In contrast, the agent of bovine spongiform encephalopathy (BSE) is extraordinarily resistant to inactivation by physical or chemical treatments. Equipment can be decontaminated by holding it in moist heat (~140°C) at alkaline pH. Methods for nondestructive decontamination of the BSE agent in livestock products, medicines and feedstuffs are needed. However, BSE is not contagious through contact or aerosols, and infected carcasses can be moved to central sites for disposal. Furthermore, controls are in place to ensure that an attack is unlikely to result in the need to slaughter large numbers of BSE-infected animals. Disposal is by incineration or alkaline digestion with heat and pressure (DNV 1997).

Decontamination of seeds and combines, trucks, or other field or handling equipment with chemical fumigation adds to the economic and environmental costs of controlling Karnal bunt and other plant pathogens or pests. Methyl bromide, one of the standard and preferred methods for fumigation of soil and containers, will no longer be legal for use after 2005 in developed countries and after 2010 in developing countries, because of an international agreement in response to evidence that it affects the ozone layer. Treatments, such as the use of live steam, can be used to clean up facilities and handling equipment, but costs and damage to the equipment can make this method prohibitive. Alternative methods for such decontamination are needed (Chapter 5, Recommendation I.F.2.d).

The resistance of bacterial spores to decontamination was illustrated after the anthrax attacks in the United States during the fall of 2001. Improved understanding of the ecology of microbial spores in the soil and of the natural enemies of spores in soil, as well as the development of novel methods to inactivate soil-borne spores, are needed. It is also important to develop assays that can distinguish between live and dead agents in the environment, so that rapid assessments of the effectiveness of decontamination procedures can be made. Most antibody and PCR-based methods cannot be used for this purpose.

SOCIAL AND PSYCHOLOGICAL DIMENSIONS

The social sciences, coupled with public health, play a major role that complements the role of the biological and physical sciences in the ability of the United States to deter, prevent, thwart, respond to, and recover from an intentional biological attack on the nation. This complementary role would be especially critical when the attack is intended to destroy or raise doubts about the safety of the nation's food supply, damage trust in government, or inflict harm to the fabric of society.

In addressing psychological and social issues related to response to and recovery from (consequence management) a bioterrorist attack, Becker (2001a) identifies several key problems. By his analysis, current response and recovery approaches are insufficiently interdisciplinary, devote too little attention to social impacts such as stigma, focus too little on longer-term issues, rarely consider scenarios where the impacts are primarily psychosocial, and focus too little on fundamental overriding (macro-level) issues, such as the maintenance and reestablishment of trust.

There are several reasons to study social and psychological dimensions of agricultural bioterrorism, among them to improve our ability to profile potential perpetrators and our intelligence methods, to develop more effective education and communication approaches, to increase compliance with mitigation measures in crisis situations, and, most important, to help victims of such attacks. This section reviews some key issues on which better social science and psychological perspectives are needed.

Intelligence and Risk Profiles

There is a pressing need to study social and psychological dimensions of agricultural terrorism to improve our ability to profile potential perpetrators and use more-effective intelligence methods (Chapter 5, Recommendation I.F.1.d). The political, cultural, religious, economic, and other issues that have led to state-sponsored or state-tolerated terrorism and terrorist organizations are probably the same whether the form of terrorism is chemical, nuclear, or biological. Reports on the value of intelligence and communication in the legal and medical communities in the Tokyo subway sarin incident provide evidence of the benefits of using intelligence for continuing assessment of terrorist risk profiles (NRC 1999).

In the United States, recent experience with deliberate threats to agricultural plants and animals has involved organized groups opposed to the use of animals in research or opposed to genetic engineering of crops (ISB 2000). Although biological agents of mass destruction have not been used in these instances, the destruction of plants in research plots and the release of experimental animals to fend for themselves constitute threats nonetheless. Little is known about the potential effects on research being conducted under such a threatening environment. Recent US experience with bioterrorist acts intended to affect food safety, (for example, after the food or product has left the farm or ranch) has involved mainly domestic groups or individuals with a grudge or agenda aimed at a specific individual, company, or community. In one case, a supplier for a Purina Mills animal feed plant in Wisconsin was threatened in 1996 with a contaminated feed ingredient; officials determined that the feed products had been contaminated with chlordane and indicted an owner of a rival company (Jones 2002).

At this writing (Spring 2002), reports in the news media have concluded that the highly potent formulation of anthrax delivered through the US Postal Service was the work of a terrorist using a formulation of a well-studied domestic strain of the agent. Likewise, agricultural plants and animals could be targets of terrorists who use strains of pests or pathogens obtained directly from legitimate research laboratories or other facilities.

The Victims: Consumers, Producers, and Community

One of the most important roles of the social and behavioral sciences involves understanding and responding effectively to the psychological and social effects of agricultural bioterrorism on the victims of such attacks (Chapter 5, Recommendation I.F.1.d). The victims include farmers, ranchers, food processors, handlers, and retailers, as well as the surrounding community. Suddenly and unexpectedly, the items that they produce and sell may be subject to recall or quarantine, and they fear loss of economic sustainability. The victims could also include those who respond to attacks—veterinarians, field agronomists, and other practitioners responsible for making correct diagnoses, gathering accurate information, and containing pests or pathogens. The victims include people, communities, and businesses at the local, regional, and national levels that depend on agriculture for their livelihood. In the case of criminal acts by groups opposed to animal or biotechnology research, the victims include researchers and staff of facilities targeted for attack. Finally, the victims include society as a whole—when it is caused to doubt the safety of its food, when it sees the trauma of wholesale slaughter of livestock each day on the evening news, or when it recognizes that the threat to food security or safety is caused not by the course of natural events but by human beings.

The disruption or destruction of the economic life of any locally important plant or animal industry can have devastating regional consequences but little or no effect on consumers or society generally. For example, due almost entirely to consecutive years of severe Fusarium headblight (scab) (caused by *Fusarium graminearum* starting in 1993), net farm revenue went from a profit of $14 and $19 per acre in Northeastern North Dakota and Northwestern Minnesota, respectively, in 1992, to net losses each of the next six years in NE North Dakota and five of the next six years in NW Minnesota (Wendels 2000). This region represents the northern-most portion of the highly productive Red River Valley that extends into Manitoba where scab was also severe in the years 1993-1998. The region grows mainly hard red spring wheat, durum, and spring barley, all of which were impacted by scab. The disease caused a double hit of both lower yields—an average of 50% lower in NE North Dakota in 1993—and lower price, because of poor quality grain and contamination by the mycotoxin deoxynivalenal (DON). By 1998, the average debt:asset ratio for farms in the region had increased to 56%, causing many second and third generation farm families to sell their farms. The impact has been equally severe on the small towns, churches, and local businesses supported by the income generated by wheat from those farms. Because some 95% of the cost of bread is made up of off-farm costs (USDA-ERS 2002c), the local social and economic impacts had virtually no effect on the price of bread, and few Americans outside the affected areas are aware of the consequences of this plant disease.

There may also be populations at special risk, such as families with young children or producers who work with animals (NRC 1999; Becker 2001b). The social and psychological effects of the FMD outbreak in Great Britain on farmers, rural communities, children, and the general public were traumatic. The stresses on individuals, families, and communities are both immediate and long-term and include the uncertainty and fear of what the future may bring, distrust of government and science, isolation, loss of long-time pets, and feelings of helplessness (Becker 2001b).

Little investment has been made in research and education aimed at understanding or minimizing the psychological and social impacts of bioterrorism on those many groups of victims, whether involved in production or dependent on food and agriculture.

Public Information and Response

As illustrated in the FMD outbreak in Great Britain, conveying information and managing educational outreach with livestock producers and the public are critical elements in effective response to an outbreak. Many examples can be cited to illustrate the instinctively negative response of society to information that raises doubts about the safety of the food supply as it comes from the farm or ranch. Swartz and Strand (1981) analyzed consumer sales effects of information and a ban on sales of oysters harvested from waters contaminated by kepone in the Chesapeake Bay. Smith et al. (1988) showed the adverse effects on milk sales in Hawaii after an incident of heptachlor contamination of milk supplies. The apple industry was seriously affected by the response of consumers to a report that Alar, a chemical used to reduce the number of apples allowed to mature on the tree, was present in apple sauce used as baby food and might cause cancer. A study in the New York City market showed substantial sales losses due to shifts in consumer demand after an announcement of possible risks posed by foods with Alar residues (van Ravenswaay and Hoehn 1991).

Those studies establish that consumers respond to negative information about the safety of the food supply. What is less well understood is how they deal with conflicting information and how consumers and others can be educated about the relative risks posed by foods and food production. There is some evidence from behavioral and economic studies that negative information outweighs favorable information in consumer and producer choices concerning risks. Results of consumer studies of irradiation of food show that when consumers are presented with unfavorable and favorable information at the same time, unfavorable information dominates (Fox, Hayes and Shogren in press). Viscusi (1997) found that people place greater weight on the higher risk (worst case) when provided with information from two sources. An explanation offered is that of prospect theory (Kahneman and Tversky 1979, Thaler 1980): people tend to make choices aimed at averting losses more than at achieving potential and corresponding gains. In other words, at a given reference point, people attach more weight to monetary losses associated with switching than to similar gains. The potential for overriding the influence of unfavorable information suggests the importance of developing effective public communication and education strategies (Chapter 5, Recommendation I.E.3).

Because producer and consumer responses are likely to be crucial in the general response to and recovery from a bioterrorist attack, understanding individual behavior under conditions of stress (psychological, economic, and social) will allow more effective communication strategies to be developed. That is especially the case in the wake of a bioterrorist attack on an agriculturally important crop or farm animal that is likely to generate social and psychological effects on farmers and rural communities that depend on the crop or animal.

The public threat to life or livelihood is different in every case but there is increased awareness of some key elements of effective response. There is a potential for loss of confidence in authorities. The case of Johnson & Johnson's response to the tampering with Tylenol indicates the importance of effective communication and strong leadership (Johnson & Johnson 2002). The CEO's leadership in making public statements of information and of what the company was doing to control the problem showed that immediate and effective delivery of information and control of product can reduce large-scale damage to public confidence.

New knowledge gained from the FMD outbreak and other disasters involving livestock or crops indicates the important role of immediate and factual communication, including both public (such as USDA-provided) and private (such as producer-provided and industry-provided). During the FMD outbreak, the Internet proved effective as a source of information when travel in a local area was restricted. Given the growing importance of the Internet, there is a need to understand this mode of delivery and its potential for use and misuse (Chapter 5, Recommendations I.E.1.d and I.E.3).

During a crisis, there might be important economic incentives for taking private actions that could delay reporting of a pest or pathogen infestation or disease. Those involved in the initial detection might well be subject to economic or personal hardship (for example, through quarantine, destruction, or recall of their products), if a disease or infestation is confirmed. Becker (2001b) reports that in the UK, FMD warning signs posted on footpaths and roads were torn down on a number of occasions. The challenge is to encourage private support for the public good. Understanding how to educate producers and create appropriate incentives and compensation schemes is necessary if effective plans and strategies for control and containment of the effects of an attack are to be developed. Better understanding of how to present information, educate producers and the public, and encourage socially beneficial behavior is likely to reduce costs and make controls more effective. Incorporating information on the psychological and social dimensions will improve strategic planning and help to maintain public confidence (Chapter 5, Recommendation I.F.1.c).

RESEARCH NEEDS

In light of the foregoing discussion, the committee identified the following kinds of research as necessary to improve our ability to counter agricultural bioterrorism. As previously mentioned, R&D priority setting is a necessary part of an overall defense plan, and this list can serve as a starting point for such an effort.

Detection, Identification, and Diagnosis

- Develop reference specimens and other taxonomic information for pests or pathogens likely to be used in bioterrorist attacks against US agriculture so that rapid and accurate identification can be made after a pest or pathogen is discovered.
- Develop rapid detection, identification, and diagnostic methods that can be deployed in the field by inexperienced or non-specialist workers.
- Develop improved methods for early recognition of the presence of pests and pathogens, including biosensors and remote sensing technology.

Genomics, Genome Sequencing, and Transgenics

- Develop better methods for using genomic information, information technology, and molecular "fingerprinting" to trace threat agents to their origin.
- Develop primers and probes for real-time PCR that can identify relevant threat agents and develop a database of such information.
- Encourage global efforts to obtain full-length sequence information of strains and subtypes of likely or high impact threat agents.
- Exploit genomic research and information in order to develop crop varieties that are resistant to key pests and pathogens (for example, those that are most likely to be used in a terrorist attack and to be most devastating).

Information Technology

- Develop a coordinated information technology base for agricultural bioterrorism including information on gemonic sequences for bioforensics and research; detection, identification, and diagnostics obtained by laboratory networks; epidemiological analyses; outbreak control information; a directory for experts in related subject matter; and relevant data from research and development projects, field stations, regional agriculture laboratories, and agriculture professionals.

Control Technology

- Develop epidemiological models for key threat agents so that plans can be put into place for containment of introduced organisms and for the integrated management of infestations or diseases that cannot be eradicated.
- Improve understanding of the ecological reservoirs of agricultural pests and pathogens, including the behavior and ecology of organisms that serve as vectors for such agents.
- Develop or revise specific control (management or eradication) plans for key pests or pathogens most likely to be used in bioterrorist attacks against US

agriculture. Test key control programs, whether the goal is eradication or management, in the geographic regions of origin of pests or pathogens. Such tests have the double benefit of assisting other countries in developing effective control technology and often ensuring that we have the knowledge and ability to control key pests and pathogens introduced into the United States in a bioterrorism event.

- Enhance the ability to use selective pesticides, antibiotics, and other chemotherapeutic agents while reducing the negative effects of this management tactic. Selective use of pesticides and antibiotics might involve altering how, when, and where these products are used. Developing knowledge on key agents that might be used in bioterrorist attacks will be important for deterring and responding to agricultural bioterrorism.
- Develop improved vaccines for key pathogens of animals. If it is appropriate, stockpile sufficient supplies of vaccines for such animal pathogens so that they can be deployed rapidly.
- Explore technology that would be needed to support timely vaccine development and production after discovery of an outbreak. Develop newer technologies (such as DNA, vectored and replicon systems) that could be conditionally or fully licensed as generic systems now and then pushed to production rapidly (in weeks) if they are needed. Support basic immunology and vaccine R&D on the new technologies.
- Develop information on key natural enemies that could be introduced into classical biological control programs directed against agricultural bioterrorism agents. This would shorten the time that such a program would take and allow informed risk assessments of proposed introductions of natural enemies.
- Develop more efficient and effective methods for large-scale disinfection of soils, equipment, and facilities.

Social Sciences

- Support research to understand the perpetrators of agricultural bioterrorism as one approach to deter, prevent, or thwart bioterrorist attacks.
- Enhance understanding of the concerns of food producers and the public. The resulting information can be used to develop more effective and responsive communication strategies.
- Support research on the social and psychological effects of bioterrorism, including identification of high risk groups and ways of assisting affected individuals, families and communities.
- Investigate methods of educating producers and consumers about plant and animal diseases and infestations.

5

Strengthening the Nation's Defense Against Agricultural Bioterrorism

The charge to the committee was to evaluate the ability of the United States to deter, prevent, detect, thwart, respond to, and recover from intentional biological attacks on the nation at the live-animal and live-plant stage of food and fiber production (Box 1-2). Through its analysis of the US agricultural defense system in Chapter 2, exploration of the specific agents and lessons learned in Chapter 3, and the identification of scientific research and technology needs in Chapter 4, the committee came to the following key conclusions: 1) the United States is vulnerable to bioterrorism directed against agriculture, 2) the nation has inadequate plans to deal with it, 3) the current US system is designed for defense against unintentional biological threats to agricultural plants and animals, and 4) although strengthening the existing system is a resource-efficient and effective part of the response to bioterrorism, it is not sufficient. The committee recommends a concerted effort on the part of the US government to develop a comprehensive plan to counter agricultural bioterrorism. Details of the recommended plan are presented below with other specific findings and conclusions that support the need for its development.[1] The committee believes that the suggested plan can make the system stronger, more science-based, and ultimately, make the nation less vulnerable to intentional biological threats.

The committee recognizes that this report portrays a snapshot of the issues as they existed between August and December 2001. As it was being prepared, many changes to the US defense system were being proposed, and some imple-

[1]The body of this report lends further support. Supporting sections in the body of the report are cross-referenced in this chapter. Likewise, the numbered findings, conclusions, and recommendations in this chapter are cross-referenced in supporting sections in Chapters 2, 3, and 4.

mented. The committee tried to analyze and incorporate the changes up until the spring of 2002. The committee hopes that its work will help to inform the ongoing debate and encourage change to prepare the nation better for defending agriculture.

KEY FINDINGS AND CONCLUSIONS

This section presents the committee's findings and conclusions (summarized in Box 5-1) that support the recommendations in the following section.

Finding I. The United States is vulnerable to bioterrorism directed against agriculture.

Biological threat agents that vary in the nature and extent of their potential impacts are widely available for intentional introduction into agricultural plants and animals and pose a substantial threat to US agriculture. Technical sophistication would not be necessary for attacks with some threat agents. The current US plan and system were designed to defend against and respond to unintentional introductions and are useful, but are inadequate for intentional events (Chapter 2).

Although an attack with such agents is highly unlikely to result in famine or malnutrition, the damage that could occur includes major direct and indirect costs to the agricultural and national economy, adverse public-health effects, loss of public confidence in the food system and in public officials, and widespread public concern and confusion. In this report, the committee provides examples of economic costs of agricultural infestations or diseases (Table 5-1). Costs for various scenarios range from millions of dollars to tens of billions of dollars, and vary by value of crop or animal, costs of eradication or control, and likely trade effects, among other factors. Economic and social burdens would be particularly evident in rural communities. The global nature of our agriculture and food system makes the United States vulnerable to attacks introduced at either foreign or domestic sites. Because of trade and market impacts and the public's unique sensitivity about food, an outbreak that affects only a few animals or crop fields—or in some cases even a hoax—could have major economic, social, and psychological consequences. In some cases, recent industry concentration and specialization make US agriculture even more vulnerable to attack (e.g., widespread movement of animals and seeds, or rearing of large numbers of livestock in proximity to each other).

The following findings and conclusions support the committee's overarching tenet that US agriculture is vulnerable to bioterrorism.

Finding I.A. Intentional introductions of pests and pathogens may differ substantially from unintentional introductions.

This report points out that there are weaknesses in the US agricultural defense against unintentional biological threats, but it is imperative to recognize that

BOX 5-1
Findings and Conclusions

I. The United States is vulnerable to bioterrorism directed against agriculture.

A. Intentional introductions of pests and pathogens may differ substantially from unintentional introductions.
B. The nation has inadequate plans to deal with agricultural bioterrorism.

1. As of spring 2002, no publicly available, in-depth, interagency or interdepartmental national plan had been formulated for defense against the intentional introduction of biological agents directed at agriculture.
2. The adverse effects of bioterrorism agents on wildlife have been little considered.

C. There are important gaps in our knowledge of foreign-plant and foreign-animal pests and pathogens. These gaps reduce the reliability and timeliness of risk assessments and risk-management decisions.
D. The current inspection and exclusion program at the US borders, in which only small proportions of people and goods entering the United States are inspected, is inadequate for countering the threat of agricultural bioterrorism.
E. Our ability to rapidly detect and identify most plant pests and pathogens and some animal pests and pathogens soon after introduction is inadequate. This allows them to spread, results in greater damage, and makes it more expensive or impossible to respond with eradication.
F. A large-scale multifocal attack on agriculture could not be responded to or controlled adequately or quickly and would overwhelm existing laboratory and field resources.
G. It is not feasible to be specifically prepared or have all the scientific tools for every contingency or threat to agriculture.
H. Although the nation's fundamental science, research, and education infrastructure (academic, industrial, and government) is in place and functional, preparing the nation for agricultural bioterrorism requires special efforts and support of the infrastructure.
I. There is a need to enhance the basic understanding of threat agents so as to develop new and exploit emerging technologies for rapid detection, identification, prophylaxis, and control.

TABLE 5-1 Examples of Costs of Outbreaks of Some Plant and Animal Diseases or Infestations

Agent and target	Cost	Geographic area	Year	Reference
Chlordane in feed (intentional introduction)	$4 million	Wisconsin	1998	Neher 1999
FMD in cattle	$6-30 billion	UK	2001	Ferguson 2001 Becker 2001a
FMD in pigs	$6 billion for slaughter	Taiwan	1997	Beard and Mason 2000
BSE in cattle	$2.5 billion (direct compensation to farmers for animals over 30 months)	UK	1996-2002	DEFRA 2002
BSE in cattle if it were to occur in US	$3.7 billion for eradication	US	1997 estimate	FDA 1997
Mediterranean Fruit Fly eradication	$400 million	California and Florida	1908-2000	CDFA 2002
Avian influenza in poultry	$65 million	Northeast US	1983-1984	USDA-APHIS 2001c
African Swine Fever in pigs If it were to occur in Texas	$34 million	Texas	1994 estimate	Rendelman 1994
Screwworm eradication	$54 million	Cuba	1958	Meadows 1985
Citrus Canker in citrus trees	$200 million	500 mi^2 in Florida	mid 1990s to 2001	Schubert 2001
Unwanted transgene in food supply	$90 million recall costs	US	2000	Harl et al. 2002
Karnal bunt in wheat	$27 million (direct and indirect losses)	4 Counties of north Texas	2001	Bevers et al. 2001

merely fixing these weaknesses will not adequately prepare the United States against intentional attacks, for several reasons (Chapter 2) (Sequeira 1999). The perpetrators will have the advantage of selecting unanticipated and covert means, the time of introduction, introductions into remote areas, multiple introductions of agents, and simultaneous release of multiple species. Intentional introductions permit an increased probability of survival of a pest or pathogen in transit, the use of highly virulent strains and high concentrations of inoculum, and precise timing of release to coincide with maximal colonization potential. Attacks also differ from unintentional introductions in that the perpetrator is able to target susceptible production areas and natural environments and can include sabotage of laboratory and field testing resources.

Finding I.B. The nation has inadequate plans to deal with agricultural bioterrorism.

Finding I.B.1. As of spring 2002, no publicly available, in-depth, interagency or interdepartmental national plan had been formulated for defense against the intentional introduction of biological agents directed at agriculture.

The US Department of Agriculture (USDA) Animal and Plant Health Inspection Service (APHIS) has several emergency plans for dealing with unintentional introductions of plant and animal pests and pathogens, but the committee could not find any publicly available interdepartmental national plan designed for defense against an intentional introduction of a plant or animal pest or pathogen (Chapter 2).[2] Likewise, there is no accepted interagency threat list for agricultural bioterrorism that can be used for developing a defense response plan that includes coordination among partners, laboratory strategy, public information, and determination of research needs. Coordination within and among key federal agencies and coordination of federal agencies with state and local agencies and private industry appears to be insufficient for effectively deterring, preventing, detecting, responding to, and recovering from agricultural threats (Chapter 2). For some situations, the immediate response by front-line personnel will not differ in intentional versus unintentional introductions. However, in many cases, it will, and specific planning is justified. A well-organized plan is crucial for mitigating the adverse impact of such attacks.

Finally, although the committee finds that preparation for agricultural bioterrorism enhances our ability to maintain national agricultural health and productivity, in converse, preparations only for unintentional outbreaks are not adequate for agricultural bioterrorism.

---

[2]The committee recognizes that such plans might be classified and could not be openly discussed with the committee. The committee has received no indication, however, that such plans exist.

Finding I.B.2. The adverse effects of bioterrorism agents on wildlife have been little considered.

Bioterrorism agents could have long-term and detrimental effects on biodiversity and ecosystems. A consideration of effects on the environment and wildlife, which can act as reservoirs for diseases and infestations of livestock and crops, is necessary in planning for defense against agricultural bioterrorism. Most attention is focused on direct introduction of a pest or pathogen into crops or livestock. However, an infection could be first established in wildlife where it might remain undetected for a long period, because wild plants and animals are not monitored as closely as commercially important organisms and often carry the agent without signs. It would be difficult to control such an attack, and reinfection from wild animals would pose a continual threat.

Finding I.C. There are important gaps in our knowledge of foreign-plant and foreign-animal pests and pathogens. These gaps reduce the reliability and timeliness of risk assessments and risk-management decisions.

The basic biology of many common agricultural pests and pathogens is not well understood (Chapter 4), and improved surveillance overseas is needed (Chapter 2). Risk assessment is essential for planning and coordinating counterterrorist activities. Assessments need to be based on global intelligence concerning potential terrorist activities and on continuing international surveillance for new and emerging diseases and infestations, and they need to be regularly updated. Risk assessments (conducted jointly by agencies responsible for intelligence and agriculture) of the most probable means for acquiring, introducing, and disseminating each of the major intentional threat agents are needed.

Finding I.D. The current inspection and exclusion program at the US borders, in which only small proportions of people and goods entering the United States are inspected, is inadequate for countering the threat of agricultural bioterrorism.

Current methods for interdicting pests and pathogens at borders, such as oral and written declarations and x-ray screening, are not targeted at intentional biological threats (Chapter 2). In addition, although all travelers, conveyances, cargo, mail, and other articles arriving in the United States from foreign origins are subject to inspection, only a relatively small percentage are actually inspected by APHIS. Furthermore, inspections focus on prohibited products (that is, those with high risk for unintentional introduction of biological threats) rather than on deceptively packaged threat agents which are likely in intentional introductions. Recognition of packaging to keep agents viable and delivery and dispersion devices that may be linked with successful use of an agent are important. New technologies and novel inspection methods are needed to improve our ability to detect key pests and pathogens and their packages and delivery systems at seaports and border crossing sites (Chapters 2 and 4). Exclusion methods must become faster, more sensitive, and more cost-effective.

Improved border controls can make unintentional introduction of foreign threat agents more difficult and thus reduce the risk of introduction somewhat, and border security is an integral and critical component of a multitactic strategy to reduce the threat of agricultural bioterrorism. However, it cannot be relied on to prevent intentional introduction. For example, APHIS Plant Protection and Quarantine (PPQ) relies heavily on port-of-entry pest-interception data to identify potential foreign-plant threat agents. However, a foreign pest or pathogen that is readily available to terrorists might not present a substantial threat through unintentional pathways; in that case, it would not be on the radar screen of border inspectors (Chapter 2). In addition, committed, sophisticated terrorists are likely to devise ways to circumvent such controls. Thus, our national capacity for rapid detection and response after introduction is of paramount importance in countering agricultural bioterrorism.

Likewise, methods of preventing or deterring threat agents of domestic origin are needed and could include increased security of US laboratories to prevent access to agents and better on-farm security to prevent transfer of threat agents to the target. On-farm security will be difficult, however, especially for field crops, because they are generally grown over large expanses and are not closely monitored.

Finding I.E. Our ability to rapidly detect and identify most plant pests and pathogens and some animal pests and pathogens soon after introduction is inadequate. This allows them to spread, results in greater damage, and makes it more expensive or impossible to respond with eradication.

Early detection, identification, and diagnosis are extremely important in limiting the damage caused by pests and pathogens (Chapters 2 and 4). It is generally less expensive to detect, prevent the spread of, and eradicate agricultural pests and pathogens rapidly than to deal with the pest or pathogen once it is widespread or established, when we may have to learn to live with it. If introduced agents cannot be eradicated, there will be a continual threat to the sustainability of a sector of US agriculture because of increased costs for detection, suppression, and management, and because of decreased productivity and losses in international trade.

Time is of the essence in minimizing damage. The lag between pest or pathogen establishment and detection varies widely and directly affects the ability of APHIS and other groups and individuals to respond effectively. Current programs rely on initially detecting the presence of pests and pathogens by observing the health of animals or plants (Chapter 2), but the lag from initial infection to display of clinical signs ranges from, for example, 2-14 days for foot and mouth disease (FMD) to years for bovine spongiform encephalopathy (BSE, or "mad cow disease"). By the time diseased or infested plants or animals are noticed, the pest or pathogen has the potential to spread well beyond the points of its initial introduction. In addition, no universal system is in place for screening, identifying and reporting plant pests or pathogens found by passive surveillance (surveillance performed not under a formal program, but rather informally through growers, extension agents, and so on). The ability to detect and identify agricultural pests and pathogens varies from state to state. For plant pathogens and pests, the system is further hindered by underfunding and understaffing of state diagnostic laboratories. Although suspicions of intentional releases might now be heightened in light of the tragedy of September 11, 2001, there is still a need to build the infrastructure and communication systems to act on these suspicions.

Finding I.F. A large-scale multifocal attack on agriculture could not be responded to or controlled adequately or quickly and would overwhelm existing laboratory and field resources.

Current laboratory and field resources are often strained by naturally occurring outbreaks. For example, university plant clinics play important roles in early detection of arthropod pests, plant parasitic nematodes, and plant pathogens, but they are typically understaffed and lack resources (Chapter 2). There is seldom time or resources to perform molecular or biochemical assays on routine samples. The laboratories commonly handle hundreds or even thousands of specimens each year from home gardeners, creating the potential for backlogs in responding to requests for diagnoses by growers and agribusinesses.

Those laboratories and the PPQ Rapid Response Teams would probably not be adequate to address a multifocal intentional introduction of a biological threat agent directed at plants (Chapter 2). Similar problems might exist for APHIS Veterinary Services emergency response teams and federal and state animal-disease diagnostic laboratories, depending on the extent of a single outbreak or multiple introductions of threat agents. A surge capacity that can draw on industry and academic resources in times of crisis is needed. Response plans need to be developed for the highest-priority categories of plant and animal pests and pathogens to test the level of response preparedness. A "bioterrorism rapid response" strategy for agriculture is needed (Chapter 2). Response planning needs to be better coordinated with the foreign and domestic intelligence communities (Chapter 2).

Finding I.G. It is not feasible to be specifically prepared or have all the scientific tools for every contingency or threat to agriculture.

Although recent legislation has provided additional funding for combating agricultural bioterrorism (Chapter 2), resources are still low, and numerous scenarios are conceivable (Chapter 4). In addition, a terrorist might have the ability to imagine a scenario that is inconceivable to the planners (Chapter 2). Therefore, plans can focus only on a subset of the possible threats. The subset should cover diverse agent and target species that seem plausible to biologists, agricultural specialists, and the intelligence community. The subset of agent-target combinations used for planning can help to prepare for other possible scenarios.

Finding I.H. Although the nation's fundamental science, research, and education infrastructure (academic, industrial, and government) is in place and functional, preparing the nation for agricultural bioterrorism requires special efforts and support of the infrastructure.

Several research efforts are directed against naturally occurring plant and animal pests and pathogens (Chapter 4). Information on and strategies aimed at protection of plants and animals against naturally occurring or accidentally introduced (unintentionally introduced) pests and pathogens will help but are insufficient for protection against deliberate introductions, as these introductions can differ widely (see Finding I.A). Future research on plant and animal pests and pathogens needs to encompass the unique challenges presented by deliberate introductions. For example, genetically engineered pest or pathogen strains could be used in a deliberate attack, and special considerations for detection and control are needed to account for them. Special efforts are also needed in the recruitment, training, and education of scientists and other experts in order to perform the research and analysis necessary for agricultural defense. In addition, effective and credible communication and public-information strategies are critical in preparations for response to intentional attacks, as are strategies for addressing the social and psychological effects of bioterrorism on communities (Chapter 4).

Finding I.I. There is a need to enhance the basic understanding of threat agents so as to develop new and exploit emerging technologies for rapid detection, identification, prophylaxis, and control.

The ability to prevent, detect, respond to, and recover from agricultural diseases or infestations depends on the availability of sound information on the biology of the introduced organisms, susceptibility or vulnerability attributes of agricultural plants and animals to the potential biological threats, and the molecular pathogenesis of the resulting diseases (Chapter 4). Large gaps exist in the fundamental knowledge of even the most common agricultural pests and pathogens and their effects. Additional information would underpin the development of new technologies for early detection, identification, prophylaxis, and control.

KEY RECOMMENDATION: A COMPREHENSIVE PLAN

The elements of the committee's main recommendation are summarized in Box 5-2. More details and references to supporting sections in the report appear below. The committee recognizes that a system for defense against unintentional threats is in place. Therefore, for the recommended plan, it proposes to build on

BOX 5-2
Recommendations for a Comprehensive Plan to Counter Agricultural Bioterrorism

I. The US government should establish a comprehensive plan to respond to the threat of agricultural bioterrorism that does the following:

A. Integrates elements of deterrence, prevention, detection, response, and recovery.
B. Includes domestic and international strategies for recognition, prevention, and control.
C. Defines legal and jurisdictional authority and lead roles at the federal, state, local, and private levels and includes specifications for interagency cooperation.
D. Defines a categorical priority list of threat agents for planning.
E. Establishes operational capacity in:
1. Surveillance, laboratory diagnosis, and electronic reporting of threat agents with
a. Domestic and international surveillance for selected agricultural bioterrorism agents.
b. A laboratory-response network for the detection, identification, and specific diagnosis of pest infestations and plant and animal diseases that might result from agricultural bioterrorism.
c. Accelerated evaluation, validation, and adoption of emerging technologies for rapid detection and identification of threat agents, including agents modified by recombinant-DNA methods.
d. A nationwide system (for example, analogous to the Centers for Disease Control and Prevention's Health Alert Network) for communication, data management and analysis, information dissemination, and education.
2. Response to, cleanup of, and recovery from a bioterrorism event with
a. Development and exercising of model scenarios of bioterrorism attacks using pathway analysis.
b. Development of appropriate eradication and management plans in advance of attack for selected agricultural pests and pathogens, including chemotherapeutics, vaccination, or plant- or animal-breeding programs.
c. Coordination among agricultural, wildlife, public health, human services, emergency management, intelligence, and law enforcement programs.
d. International cooperation and outreach to assist other countries in managing or eradicating agricultural threat agents.

continues

3. Public information, education and outreach with
   a. Establishment and training of credible spokespersons for classes of threat agents.
   b. Development of specific media and public information for threat agents and outbreaks, including Internet-based information and training programs.
   c. Training of local officials in mass-media and public-information responses.
   d. A comprehensive educational program for the agricultural community at large to increase recognition of infestation and disease and improve response.

F. Establishes a comprehensive science and technology base to:
   1. Increase basic understanding of
      a. Biology and epidemiology (including molecular epidemiology) of exotic agents in native and new environments.
      b. Pathogenesis in animals and plants (including through microbial genomics).
      c. Social and psychological impacts of agricultural terrorism.
      d. Perpetrators of terrorism, as one approach to deter, prevent, or thwart bioterrorist attacks.
   2. Develop
      a. Interdiction, detection, diagnostic, and identification tools.
      b. Bioinformatics and information technology.
      c. Prophylaxis and therapeutics.
      d. Control and eradication strategies and tools.

G. Establishes a public-private advisory council on agricultural bioterrorism at the level of the Secretary of the US Department of Agriculture.

the existing infrastructure (for example, APHIS emergency response) with special planning, research, and programs for intentional threats.

The committee has recommended a plan that will certainly require significant additional resources for the federal, state, and local agencies and organizations involved in implementation. The committee recognizes that decisions to direct resources to countering agricultural bioterrorism will need to be weighed against decisions to direct resources to other societal problems, including other forms of terrorism. However, it was beyond the committee's scope to perform cost-benefit analyses and pass judgement on the relative importance of agricultural bioterrorism.

Recommendation I. The US government should develop a comprehensive plan to respond to the threat of agricultural bioterrorism that does the following:

- **Integrates elements of deterrence, prevention, detection, response, and recovery (*Recommendation I.A*).**

The committee reviewed various points where intervention to deter, prevent, thwart, detect, respond, and recover can take place (Chapter 2).

It is unlikely that intervention at any one point can provide acceptable security against intentional introductions (e.g., see Finding I.D). A series of interventions is more likely to reduce the threat of introduction or mitigate its consequences after introduction, so all elements for countering agricultural bioterrorism need to be incorporated into the plan.

- **Includes domestic and international strategies for recognition, prevention, and control (*Recommendation I.B*).**

The committee reviewed global and domestic strategies for recognizing potential biological threats to plants and animals, preventing them from occurring, and ultimately controlling them (Chapter 2). Activities include those in foreign countries (off shore), such as stationing US personnel for disease or infestation surveillance, Office International des Epizooties animal disease reporting, and those in the United States, such as inspection at the US border. However, there are no concerted national or international strategies to counter intentional introductions of plant and animal pests or pathogens.

For example, no clearly defined system exists for screening, identifying, and reporting to PPQ plant pests and pathogens detected by ad hoc surveillance (Chapter 2). Current identification and reporting procedures rely heavily on the knowledge, awareness, and voluntary action of the involved parties. Therefore, the committee recommends that domestic and international strategies for recognition, prevention, and control of plant and animal pests and pathogens be developed as part of an overall defense plan.

- **Defines legal and jurisdictional authority and lead roles at the federal, state, local, and private levels and includes specifications for interagency cooperation (*Recommendation I.C*).**

A federal plan for unintentional introductions of animal disease exists and is summarized in Table 2-1. However, as of spring 2002, there was no publicly available, in-depth, interagency or interdepartmental national plan had been formulated for defense against the intentional introduction of biological agents directed at agriculture. (see Finding I.B.1). It is unclear how the federal agencies would cooperate with each other and other sectors during an attack (Chapter 2). Therefore, the committee recommends that an overall plan for countering bioterrorism include delineation of the roles and coordination of activities of USDA with key partners, for example, the Federal Emergency Management Agency, defense, intelligence, public health, law enforcement, state and local agencies, private industry, trade associations, universities, and key global partners. Counterterrorism planning should include an advance decision as to whether forensic or economic concerns will have priority in responding to an attack. Planning can begin with improved organization and commitment of existing resources, but implementation will probably require additional resources. The plan should include an implementation strategy which assigns responsibilities to agencies and organizations and holds them accountable for its development.

The inclusion of state agencies in this plan is particularly important. The initial response to the introduction of an agricultural pest or pathogen will be at the local and state level. Federal response may be days or weeks after the initial case is identified. State animal health and plant health regulatory agencies must be key players in planning and execution of strategies to prevent or respond to biological terrorism.

- **Defines a categorical priority list of threat agents for planning (*Recommendation I.D*).**

The current system is designed to deal with unintentional threats and needs to be evaluated and redesigned to respond appropriately to intentional events. The theoretical list of intentional biological threats to agriculture is broad, and it is not feasible to be specifically prepared or have all the scientific tools for every possible threat. Although several lists of biological threats to agricultural plants and animals exist, they have been based on a different set of criteria, depending on the interests of the developers and the lists' intended uses (Chapter 2). Therefore, the committee recommends that a short, interagency consensus threat list be developed and specific planning conducted for each agent on the list to determine proper roles, resource allocations, laboratory strategy, research priorities, and so on. Planning based on a few agents with varied characteristics is likely to identify many of the issues that will be important, even if an agent ultimately confronted in an attack does not happen to be included in the agreed-on short list.

Scientists and intelligence specialists from USDA, Office of Homeland Security, Department of Defense, and other government agencies should work with industry, farmers and academe to develop this list. The list should be updated at regular intervals as new knowledge becomes available (for example, every couple of years). The committee recognizes that it might be in the best interest of national security to keep this list classified or confidential (that is, not available to the general public or potential terrorists). Risk analysis and epidemiological research and theory, in addition to intelligence information, should be used in selecting the threat agents. Planning for a few types of threat agents will address issues that are relevant regardless of the specific agents encountered in an attack.

- **Establishes operational capacity (*Recommendation I.E*) in:**

— Surveillance, laboratory diagnosis, electronic reporting of threat agents (*Recommendation I.E.1*) with

a. Domestic and international surveillance for selected agricultural bioterrorism agents

Information concerning agricultural diseases and infestations in foreign countries is often sketchy and not always time-sensitive (Chapter 2). Furthermore, little intelligence information is available on the activities of state parties and nongovernment groups regarding biological-weapons research and development (see Finding I.C). Therefore, the committee recommends that field and laboratory studies to define the prevalence and epidemiology of major strains and subtypes of threat agents on a global basis be enhanced to help predict where agents are most readily available and to help in tracing outbreaks to their origins after an attack.

Through enhanced international surveillance, scientists, extension agents, and other experts in all countries can be informed of each other's work and be more likely to notice shifts in the prevalence of biological agents or in R&D designed to damage agriculture. By sharing information, ideas, and programs, international scientific collaboration can fortify national systems for safeguarding plants and animals against intentional threats.

Likewise, active domestic surveillance of crops and livestock could decrease the time from initial infection to detection (see Finding I.C). Even a serious plant disease in a populated area can go undetected for a very long time. Therefore, the committee also recommends bolstering domestic surveillance.

b. A laboratory-response network for the detection, identification, and specific diagnosis of pest infestations and plant and animal diseases that might result from agricultural bioterrorism[3]

Several sample collection and analysis steps need to be coordinated to detect and identify biological agents in the environment or in their hosts accurately and reliably. Laboratories need to work together to recognize at the earliest possible time and track the prevalence and spread of biological agents on geographic and temporal scales. Otherwise, outbreaks might not be recognized rapidly and, depending on the agent, go unnoticed for long

[3]On May 30, 2002, USDA announced that out of its allocations for homeland security under the fiscal year 2002 defense-spending bill (HR 3338) 1) $20.6 million will be provided to state and university cooperators to establish a network of diagnostic laboratories for animal diseases, and 2) $4.4 million will be used to improve plant pest and disease diagnostic capabilities. These actions appear to be consistent with the committee's recommendation.

periods, such as 5-10 years for some plant pests and pathogens (Chapter 2). Time is critical for limiting the impact of an attack, and for recognizing that the event was intentional and identifying the terrorist. Without appropriate laboratory capabilities, the severity or extent of a threat can be misjudged, and this can lead to greater agricultural and human-health impacts.

Communication and coordination among laboratories are vital for determining appropriate and rapid response measures. As of spring 2002 there was no publicly available, in-depth, interagency or interdepartmental national plan had been formulated for defense against the intentional introduction of biological agents directed at agriculture. In the event of a large-scale or multifocal attack or one using agents that persist in the environment or spread rapidly, existing laboratory systems and resources would be insufficient for minimizing the spread and impact. The United States needs a laboratory network that can rapidly detect, identify, diagnose, report, and respond to agricultural bioterrorism threats. In Chapter 4, the committee discusses an existing laboratory network for human-health threats. Given the experience with the system and its proven ability to function in a crisis, the committee recommends that this framework be used as a starting point for developing a laboratory-response network for agriculture.

This network should be capable of integrating diagnosis and response information (for example, protocols for response actions), research information (for example, genomic sequences), and survey information (for example, recognition of strange pest or pathogen occurrences). Subnetworks should be designed to effectively manage the various components and roles of the network. The network should support everything from field testing to sophisticated confirmation of identification to biological forensics and could build on existing federal, commercial, university, state extension, and regional facilities. It must be capable of performing molecular analyses, be continuously upgraded, and be connected by an electronic laboratory reporting system. The network should also provide a surge capacity. Laboratories for agricultural, wildlife, and public-health diagnostics, the Federal Bureau of Investigation, and local law-enforcement laboratories should be incorporated into the system. Two subnetworks, one each for plants and animals, should be developed. The diagnostic results in the databases should be screened continuously for possible newly emerging or unusual patterns of disease and infestation. The network should also be used to identify laboratory capabilities and expertise.

The committee is not recommending the creation of new government institutions for this network, but rather consolidation, expansion, and improvement of existing infrastructure. For animals, the committee recommends bolstering existing federal (APHIS National Veterinary Services Laboratories at Ames, IA and Plum Island, NY), state, local, and university laboratories and integrating them into a comprehensive network, as described above. For plants, the committee recommends establishing a lead federal plant diagnostic lab (perhaps one that already exists through ARS or APHIS) which would take charge of developing, validating and standardizing diagnostic tools; maintaining reference collections of pathogens, reagents, and databases; training personnel; and providing leadership in bringing the existing plant diagnostic labs into a network of sharing information, technology, and ideas. Other reports have suggested the creation of a new institution for plant disease that mimics the Centers for Disease Control (CDC) for human disease (APS 2002).

c. Accelerated evaluation, validation, and adoption of emerging technologies for rapid detection and identification of threat agents, including agents modified by recombinant-DNA methods

Early detection and diagnosis are pivotal for limiting the extent of an outbreak, whether it is natural or intentional (Chapter 2). Technology must be rapid, field-deployable, accurate, and sensitive and should be inexpensive and require little training for use. There have been substantial advances in technology for detection and diagnosis of microbial diseases, but they have often not included agriculturally important pathogens, nor have they been inexpensive or field-deployable (Chapter 4). Therefore, the committee recommends a large expansion in detection and rapid diagnostic capabilities for high-priority pests and pathogens that are likely to be used in intentional attacks against US agriculture.

In some cases, available detection and diagnostic tests, such as polymerase chain reaction tests for FMD, are not used, because they have not been validated and accepted by the international community. Therefore, the committee recommends a national- and ideally international-mechanism to promote the testing, evaluation, and validation of emerging detection and diagnostic tests. The mechanism should be integrated into the laboratory-response network described above.

d. A nationwide system (for example, analogous to the Centers for Disease Control and Prevention's Health Alert Network) for communication, data management and analysis, information dissemination, education, and public relations

Exchanging information during crisis situations is essential for coordinating responses, mobilizing resources, preventing inappropriate actions, and ultimately minimizing impacts. Federal, state, and local agencies, industry, producers, and academic scientists need to communicate effectively with each other and be informed on a real-time basis so that threat agents can be intercepted, destroyed, or at least controlled as soon as possible. Recent test exercises involving simulated release of bioterrorism agents highlight the lack of communication among agencies and partners as an important deficiency (Chapter 2).

Honest and effective communication is also vital in maintaining public confidence and enlisting the public as a partner in mitigation measures, such as quarantines (Chapter 4) and destruction of crops, and minimizing exposure to the threat agents. For agriculture, it will be important to help the public understand the extent and severity of threats, some of which could turn out to be hoaxes, so that economic, social and psychological impacts can be reduced.

There is no national communication, data-management, and information system for agricultural bioterrorism. The committee recommends that one be established. The committee recommends patterning an agricultural system after one being designed for human-health threats, CDC's Health Alert Network (HAN). That network is a nationwide, integrated information and communication system that serves as a platform for distribution of health alerts, dissemination of prevention guidelines and other information, distance learning, national disease surveillance, and electronic laboratory reporting, as well as for CDC's bioterrorism and related initiatives to strengthen preparedness at the local and state levels.

The committee did not analyze the potential effectiveness of this network in comparison with other alternatives, but recommends the approach for agricultural bioterrorism to take advantage of current efforts and plans already under way for human health.

The recommended system should include the development of electronic laboratory reporting, for example, the National Electronic Disease Surveillance System; epidemiological analyses; outbreak-control information; a directory of experts in related fields; and relevant data from R&D projects, field stations, regional agriculture laboratories, and agriculture professionals. The information system should be consolidated with the laboratory network suggested above (see Recommendation I.E.1.b) and interactive with the existing CDC system for human health threats.

— Response to, cleanup of, and recovery from a bioterrorism event (*Recommendation I.E.2*) with

a. Development and exercising of model scenarios of bioterrorism attacks using pathway analysis

Knowledge of the overall system for response to and recovery from an agricultural disease or infestation is vital for knowing which elements to test further and fortify. The committee recommends that pathway analyses[4] for a set of potential threat agents, such as a few, be developed to illuminate weaknesses and correct them before an attack occurs. The set of agents used for this purpose should be similar to or identical with the one used for interagency planning (see Recommendation I.D). Each analysis should involve a description of the strengths and weaknesses of the system for the specific pest or pathogen and should propose points of intervention or risk mitigation. The committee recommends that model scenarios be exercised with the draft pathways so that important information about strengths and deficiencies in the defense system can be gleaned (Chapter 2). Following the exercises, the results should be analyzed and reported to high level officials, so that resources can be properly allocated to address the system deficiencies.

b. Development of appropriate eradication and management plans in advance of attack for selected agricultural pests and pathogens, including chemotherapeutics, vaccination, or plant- or animal-breeding programs

During a biological attack, timing and decisiveness are crucial. Delays in implementation of eradication or management plans allow further spread of diseases or infestations, economic damage, and harm to human, plant, or animal health (Chapters 2 and 4). For example, during the recent outbreak of citrus canker, eradication strategies had not been well thought out, leading to confusion, noncompliance, and ultimately further spread of the disease. Therefore, the committee recommends having such plans for selected threat agents (see Recommendation I.D) ready before an attack occurs to assist in smooth implementation in the event of one. Plans should be validated, either through simulation models or in regions of the world where a pest or pathogen is endemic (which will also benefit the host country). Where eradication efforts are unlikely to be successful or cannot be developed readily, plans should be made to develop management tactics (such as pesticides, host-plant resistance, chemotherapeutics, and vaccines) for the most serious target pests and pathogens. Strategies for management should be based on the epidemiology of the disease and the ecology of the pest (see Recommendation I.F.1.a). Planning should be based on the advanced identification of short and longer-term impacts and recovery needs.

c. Coordination among agricultural, wildlife, public health, human services, emergency management, intelligence, and law enforcement programs

---

[4]*Pathway analysis*, a common term in risk assessment, in this context signifies an exercise to outline or model the avenues through which a biological agent can be introduced and disseminated, propagate and spread, and ultimately cause disease or infestation. Mitigation measures are based on the most important control points, as determined by the analysis.

Previous responses to agricultural diseases or infestations have highlighted communication problems among various professional sectors. For example, during the West Nile virus outbreaks in 1999, wildlife experts had trouble in connecting with the public-health community and vice versa. In addition, USDA's current published emergency-response plans are not designed for intentional threats (Chapter 2) and therefore do not factor in the role of the intelligence communities. It is unclear in some cases whether eradication and management or investigation and law enforcement would take priority. Therefore, the committee recommends that response plans for various biological threats to agriculture define clearly the roles and priorities of the agricultural, wildlife, public-health, intelligence, and law-enforcement communities.

d. International cooperation and outreach to assist other countries in managing or eradicating agricultural threat agents

As discussed above (see Recommendation I.B), international cooperation is important for the control, management, and eradication of disease or infestation. A potential threat to the United States will always exist if a threat agent can be found in another country. The committee therefore recommends an effort to assist in controlling agricultural threat agents that occur naturally in other countries.

For example, the committee recommends that the United States be involved in organizing international meetings on FMD to identify the research needed to develop vaccines that could be used to control the disease without restricting international trade. The committee suggests that the meetings also be used to help to define the other biological, social, and financial constraints on global control and eventual eradication of FMD and to develop long-term research programs and political strategies that would begin to address the constraints.

— **Public information, education and outreach (*Recommendation I.E.3*) with**

a. Establishment and training of credible spokespersons for classes of threat agents

b. Development of specific media and public information for threat agents and outbreaks, including Internet-based information and training programs

c. Training of local officials in mass-media and public-information responses

Credible public information and effective communication are vital for effective management of bioterrorism and for reducing an attack's consequences. As a partner in any response effort, the public should be kept well informed. Responsible agencies should provide a clear and consistent message about the nature and extent of the threat, so as to facilitate mitigation measures and to minimize economic, social, psychological, and health effects (Chapters 2 and 4). It is important for the various government agencies and other organizations to come together to decide upon a public communication strategy and speak with one voice during a crisis. The committee recommends programs that incorporate the following three elements, which are critical for effective communication: 1) an identified lead spokesperson for each class of agents, 2) a public and mass-media information plan for each agent or type of agent, and 3) training of officials in effective public communication (Box 5-3). The committee also recommends that plans and training include responses to hoaxes.

BOX 5-3
Importance of Public Information in Incidents or Threats of Terrorism

Importance

- Public education and communication are deterrents to agricultural bioterrorism.
- Public education and communication can address and relieve public and producer anxiety.
- A credible spokesperson telling the straight story is essential.
- Social and behavioral effects of agricultural bioterrorism must be considered.
- A hoax can be used as a weapon of bioterrorism. It can lead to substantial economic loss and to loss of consumer confidence in the food supply and in public officials.

Possibilities

- USDA should set up a hotline that will field all incoming questions from the general public about agricultural bioterrorism threats and refer them to experts.
- Communication should make use of Web and e-mail resources, trade associations, industry, and extension services.
- USDA should identify expert spokespersons for specific classes of agents; these experts should work with the mass media during the response to an attack.
- Communication efforts should be informed by an understanding of public concerns.

d. A comprehensive educational program for the agricultural community at large to increase recognition of infestation and disease and improve response

The committee did not find a comprehensive educational program involving cooperative extension, academic institutions, professional or scientific societies, industry representatives, and commodity groups and aimed at increasing the awareness of pests and pathogens new or foreign to US agriculture (Chapter 2). The committee recommends the development of such a program to help to improve early recognition of agricultural threats. The program should include diagnostic information and illustrative material, and it should specify whom to call in the event of a suspicious case or discovery. The program should be tailored to likely front-line personnel (such as farmers, field agronomists, elevator operators, local veterinarians, and livestock buyers) as the primary audience, but also to public and private diagnostic and research laboratories at the regional, national, and international levels to increase awareness to the greatest extent possible.

This program should also include information about biosecurity procedures, sanitary measures, mitigation strategies to prevent infection or spread of an agent, quarantine or isolation, and practices and procedures for cleaning and disinfecting.

• **Establishes a comprehensive science and technology base (*Recommendation I.F*) to:**

— **Increase basic understanding (*Recommendation I.F.1*) of**

a. Biology and epidemiology (including molecular epidemiology) of exotic agents in native and new environments

b. Pathogenesis in animals and plants

c. Social and psychological impacts of agricultural terrorism

d. Perpetrators of terrorism, as one approach to deter, prevent, or thwart bioterrorist attacks

— **Develop (*Recommendation I.F.2*)**

a. Interdiction, detection, diagnostic, and identification tools

b. Bioinformatics and information technology

c. Prophylaxis and therapeutics

d. Control and eradication strategies and tools

The dynamic nature and extent of the threats to our agricultural economy will not allow us to plan for every specific agent or contingency. However, over the long term, a sound technical base will contribute to technology development and education for both general agricultural health and agricultural-bioterrorism preparedness and deterrence. The committee recommends enhancing basic-research programs in biology and epidemiology of agricultural threat agents, pathogenesis of crop or livestock diseases, and social-science issues related to bioterrorism. Information obtained from those programs will lead to better tools for preventing, detecting, responding to, and recovering from biological attacks.

Some tools that are needed include novel inspection technologies for use at borders (Chapter 2) and more-rapid detection and identification technologies (Chapter 4). In the short term, there are needs to identify animal threat-agent outbreaks that can be affected by a vaccination program and to analyze the cost-effectiveness of developing additional vaccine stockpiles for those agents. The committee recommends that exploration and development of newer vaccine technologies (such as DNA, vectored, and replicon systems) be pursued (Chapter 4).

Genomic information and bioinformatics have important roles in bioterrorism defense. DNA-sequence information has the potential to lead to determinations of the biogeographic lineages of pathogens and pests, which can help to determine the sources of pathogens (Chapter 4). Such information can be critical when attribution is necessary for response to and recovery from an intentional biological attack. Genomic information can also lead to the development of host-plant resistance, which, like vaccination of animals, constitutes one of the best strategies for management of plant pests and pathogens (Chapter 4).

Finally, there is a pressing need to study the social and psychological dimensions of agricultural bioterrorism to improve our ability to profile potential perpetrators and use more effective intelligence methods, to develop more-effective education and crisis communication approaches, to better understand public concerns, to increase public understanding of and cooperation with mitigation measures, and most important, to help affected individuals, families, communities, and regions (Chapter 4).

The committee suggests that intelligence and law enforcement agencies are the appropriate lead agencies for understanding the social and psychological make-up of potential agricultural bioterrorists. Understanding how to educate producers and to develop appropriate incentives and compensation schemes is necessary for developing effective strategies for control and containment of the effects of an attack, and the committee suggests that USDA is the appropriate lead agency for research and programs in these areas.

Through Chapter 4, the committee identifies more-specific examples of natural and social science R&D directions that are needed to enhance our ability to defend against agricultural bioterrorism. Resources are always limited, and not all avenues can be pursued (see Finding I.G), so R&D priority-setting is a necessary part of an overall defense plan. The committee recommends that coordinated, national mechanisms for doing so be explored.

- **Establishes a public-private advisory council on agricultural bioterrorism at the level of the Secretary of the US Department of Agriculture (*Recommendation I.G*)**

Priorities of the public and private agricultural sectors might differ in a biological attack (Chapter 2). There are also needs for better and more incentives for private-sector development of technologies for countering agricultural bioterrorism that emerge from academic and government research (for example, Chapter 4). Much industrial R&D is proprietary and not available to the government via the literature or at open scientific meetings. Furthermore, industry might not know the government's needs and plans for countering agricultural bioterrorism. An advisory council designed to 1) facilitate communication between industry and government, 2) identify potentially useful technologies, and 3) explore synergies that lead to new industrial R&D would ultimately encourage the establishment of valuable cooperative programs.

For example, there is no market incentive for industry to develop advanced technologies for diagnosis of and vaccination against FMD. The US plan is to restrict FMD diagnosis to a single federal laboratory and to use vaccination only in a restricted manner and under some circumstances, even in the face of an outbreak. If plans to counter the terrorist threat of FMD are expanded as recommended in this report (laboratory-response network, rapid and forward-deployed diagnostics, global surveillance and control, improved vaccines, and increased vaccination), the proposed advisory council might identify and recommend incentives needed for industrial development of these technologies.

The interests of the advisory council would overlap with those of the existing USDA Committee on Foreign Animal and Poultry Diseases. The advisory council could potentially be a separate stand-alone committee. Alternatively, the charter of the existing USDA committee could be greatly expanded to include 1) plants, terrorism, and domestic agents as terrorist threats; 2) members who are leaders in private sector research and development; and 3) authority to consider classified and proprietary information.

The committee did not thoroughly explore alternative mechanisms to facilitate public-private cooperation. However, the status quo is not optimal, because it tends to confine both government and industry in reactive rather than proactive modes. The proactive mode is likely to provide greater deterrence or more-rapid response (and therefore less damage) in the event of an attack. Therefore, the committee recommends that a public-private advisory council be established to act as a mechanism for building relationships, exchanging information, planning cooperative programs, and improving technology transfer between public and private sectors. The committee recommends this as an important, low-cost first step through which different or larger initiatives could be proposed.

PRIORITIES

Overall, Some of the above recommendations require more development time and resources than others, and therefore, are intermediate to long term for full implementation. The committee suggests that the following recommendations can be achieved in the near term and should have priority for immediate action:

- Establish and train credible spokespersons for classes of threat agents.
- Define legal and jurisdictional authority and lead roles at the federal, state, local, and private levels. Include specifications for interagency cooperation.
- Define a categorical priority list of threat agents for planning.
- Develop and exercise model scenarios of bioterrorism attacks.

Among those recommendations that require more time for implementation, the committee suggests that the following have priority and that plans for their implementation begin immediately:

- Increase basic understanding of pathogenesis in plants and animals.
- Establish a laboratory response network for detection, identification, and diagnosis.
- Establish a nationwide system for communication, data management, and analysis.

References

Alibek, K. 1999a. The Soviet Union's anti-agricultural biological weapons. Pp. 18-19 in Food and Agricultural Security: Guarding against natural threats and terrorist attacks affecting health, national food supplies, and agricultural economics, T.W. Frazier and D.C. Richardson, eds. Ann. New York Acad. Sci. 894:232.

Alibek, K. 1999b. Biohazard. New York: Random House.

Anderson, R.M., and R.M. May. 1991. Infectious Diseases of Humans: Dynamics and Control. Oxford: Oxford Science Publications.

APS (American Phytopathological Society). 2002. American Phytopathological Society: The First Line of Defense. Presented to Congress in February 2002. [Online]. Available: http://www.apsnet.org/media/biosecurity.asp.

Beachy, R.N., S. Loesch-Fries, and N.E. Turner. 1990. Coat protein-mediated resistance against virus infection. Annu. Rev. Phytopathol. 28:451-474.

Beard, C.W., and P.W. Mason. 2000. Genetic determinants of altered virulence of Taiwanese foot-and mouth disease virus. J. Virol. 74:987-991.

Becker, S.M. 2001a. Meeting the threat of weapons of mass destruction terrorism: toward a broader conception of consequence management. Military Medicine 166(S2):13-16.

Becker, S.M. 2001b. Presentation to Committee on Biological Threats to Agricultural Plants and Animals. November 15, 2001.

Bevers, S., T.W. McAlavy, and T. Baughman. 2001. Texas A & M News Release. November 12, 2001.

Binns, M.R., J.P. Nyrop, and W. van der Werf. 2000. Sampling and Monitoring in Crop Protection. New York: CABI Publishing.

Bottrell, D.G., P. Barbosa, and F. Gould. 1998. Manipulating natural enemies by plant variety selection and modification: A realistic strategy? Annu. Rev. Entomol. 43:347-367.

Broglie, K., I. Chet, M. Holliday, R. Cressman, P. Biddle, S. Knowlton, C.J. Mauvais, and R. Broglie. 1991. Transgenic plants with enhanced resistance to the fungal pathogen *Rhizoctonia solani*. Science 254:1194-1197.

Cao, H., X. Li, and X. Dong. 1998. Generation of broad-spectrum resistance by overexpression of an essential regulatory gene in systemic acquired resistance. Proceedings of the National Academy of Science 95:6531.

Campbell, C.L., and L.V. Madden. 1990. Introduction to Plant Disease Epidemiology. New York: John Wiley & Sons.

Casida, J.E., and G.B. Quistad. 1998. Golden age of insecticide research: Past, present, or future? Annu. Rev. Entomol. 43:1-16.

CDC. 2001a. West Nile Virus Questions and Answers. [Online]. Available: http://www.cdc.gov/ncidod/dvbid/westnile/q&a.htm [2002, February 2].

CDC. 2001b. Laboratory Response Network for Bioterrorism (LRN). [Online]. Available: http://www.phppo.cdc.gov/mlp/pdf/nls/NLS1201.pdf [2001, December 20].

CDFA (California Department of Food and Agriculture). 2002. Background on Medfly Problem in California. [Online]. Available: http://134.186.235.98/pests/medfly/background.html.

Center for Non-proliferation Studies. 2001. Chemical and Biological Weapons Resource Page. [Online]. Available: http://CNS.miis.edu/research/cbw/agchron.htm.

CFIA (Canadian Food Inspection Agency). 2000. Tripartite 2000 Final Report. [Online]. Available: http://www.inspection.gc.ca/english/anima/heasan/simulation/trirepse.html.

Chan, M.S., and M.J. Jeger. 1994. An analytical model of plant virus disease dynamics with roguing and replanting. J. Appl. Ecol. 31:413-427.

Coop, L. 2000. Online IPM weather data and degree-days for pest management decisions making in Oregon. [Online]. Available: http://osu.orst.edu/Dept/IPPC/wea/.

DEFRA (Department of Environment, Food and Rural Affairs). 2002. BSE. [Online]. Available: http://www.defra.gov.uk/animalh/bse/index.html.

Dhadialla, T.S., G.R. Carlson, and D.P. Le. 1998. New insecticides with ecdysteroidal and juvenile hormone activity. Annu. Rev. Entomol. 43:545-569.

De Wit, P.J.G.M., M.H.A.J. Joosten, R. Lauge, R. Roth, R. Luderer, N. Westerink, G. Honee, M. Kooman-Gersmann, F. Laurent, R.A.L. van der Hoorn, C.F. de Jong, P. Vossen, and R. Weide. 1998. Gene for gene interactions and the role of avirulence genes in pathogenicity and race-specific resistance. Abstracts of 7th International Congress of Plant Pathology, Edinburgh, Scotland. 1:9-16.

DNV (Det Norske Veritas). 1997. Risks from BSE via Environmental Pathways for the Environment Agency. London: Det Norske Veritas Limited.

Dutton, G. 2002. Detecting biowarfare weapons. Genetic Engineering News 21(19):1.

Enserink, M. 1999. New virus fingered in Malaysian epidemic. Science 284:407-410.

Enserink, M. 2002a. Useful data but no smoking gun. Science 296:1002-1003.

Enserink, M. 2002b. TIGR begins assault on the anthrax genome. Science 295:1442-1443.

FAO. 1996. International standards for phytosanitary measures. Reference Standard. Glossary of Phytosanitary Terms. ISPM Publication No. 5, Rome, Italy: FAO.

FDA. 1997. Federal Register. Volume 62(2), January 3.

FEMA (Federal Emergency Management Agency). 1979. Robert T. Stafford Disaster and Emergency Assistance Act. 42 U.S.C. 5122.

Ferguson, N.M., C.A. Donnelly, and R.M. Anderson. 2001. Transmission intensity and impact of control policies on the foot and mouth epidemic in Great Britain. Nature 413:542-548.

Foster, S.P., and M.O. Harris. 1997. Behavioral manipulation methods for insect pest-management. Annu. Rev. Entomol. 42:123-146.

Fox, J.A., D.J. Hayes, and J.F. Shogren. 2002. Consumer preferences for food irradiation: How favorable and unfavorable descriptions affect preferences for irradiated pork in experimental auctions. Journal of Risk and Uncertainty 24(1):75-95.

Fuller, F., J. Fabiosa, and V. Premakumar. 1997. World Impacts of Foot and Mouth Disease in Taiwan. CARD Briefing Paper 97-BP 16. Ames, IA: Center for Agricultural and Rural Development, Iowa State University.

Gibson, G., and S.V. Muse. 2002. A Primer of Genome Science. Sunderland, MA: Sinauer Associates.

Gonsalves, D. 1998. Control of papaya ring spot virus in papaya; a case study. Annu. Rev. Phytopathol. 36:415-437.

Goodner, B., G. Hinkle, S. Gattung, N. Miller, M. Blanchard, B. Qurollo, B.S. Goldman, Y. Cao, M. Askenazi, C. Halling, L. Mullin, K. Houmiel, J. Gordon, M. Vaudin, O. Iartchouk, A. Epp, F. Liu, C. Wollam, M. Allinger, D. Doughty, C. Scott, C. Lappas, B. Markelz, C. Flanagan, C. Crowell, J. Gurson, C. Lomo, C. Sear, G. Strub, C. Cielo, and S. Slater. 2001. Genome sequence of the plant pathogen and biotechnology agent *Agrobacterium tumefaciens* C58. Science 294:2323-2328.

Guilbault, G.G., B. Hock, R. Schmid. 1992. A piezoelectric immunobiosensor for atrazine in drinking water. Biosensors and Bioelectronics 7(6):411-419.

Harl, N.E., R.G. Ginder, C.R. Hurburgh, and S. Moline. 2002. The StarLink Situation. Iowa Grain Quality Initiative (Revised 3/19/02). [Online]. Available:http://www.extension.iastate.edu/Pages/grain/publications/buspub/0010star.PDF.

Iowa Department of Agriculture. 2001. Media Advisory: Iowa's Foot-and-Mouth Disease Response and Recovery Plan will be tested during an exercise on July 26, 2001. [Online] Available: http://www.agriculture.state.ia.us/fmd.htm [2002, February 11].

IPPC. 2002. United Nations Food and Agriculture Organization. International Plant Protection Convention, 1998. [Online]. Available: http://www.ippc.int.

ISB (Information Systems for Biotechnology). 2000. Attacks on GE Research Facilities Heat Up. ISB News Report, February.

ISTA (International Seed Testing Association). 1993. International rules for seed testing. Seed Sci. Technology 21:supplement.

Jacobsen, B.J. 1997. Role of plant pathology in integrated pest management. Annu. Rev. Phytopathol. 35:373-391.

Jeger, M.J., F. Van den Bosch, L.V. Madden, and J. Holt. 1998. A model for analysing plant virus transmission characteristics and epidemic development. IMA Journal of Mathematics Applied in Medicine and Biology 14:1-18.

Joerger, R.D., T.M. Truby, E.R. Hendrickson, R.M. Yojng, R.C. Ebersole. 1995. Anylate detection with DNA-labeled antibodies and polymerase chain reaction. Clinical Chemistry 41:1371-1377.

Johnson and Johnson. 2002. Credo. [Online]. Available: http://www.jnj.com/who_is_jnj/cr_index.html.

Jones, R.P. 2002. Center for Non-proliferation Studies. Agro-terrorism: Chronology of CBW attacks targeting crops and livestock 1915-2000. [Online]. Available: http://cns.miis.edu/research/cbw/agchron.htm [2002, February 25].

Kahneman, D., and A. Tversky. 1979. Prospect theory: An analysis or decision under risk. Econometrica 47:263-291.

Keen, N.T., B. Staskawizc, J. Mekalanos, F. Ausubel, and R.J. Cook. 2000. Pathogens and host: The dance is the same, but couples are different. Proc. National Acad. Sciences 97:8752-8753.

Kidwell, M.G., and A. Wattman. 1998. An Important Step Forward in the Genetic Manipulation of Mosquito Vectors for Human Disease. Proc. Nat'l. Acad. Sci. 95:3349-3350.

Kilman, S. 2001. Special Report: Aftermath of Terror. Wall Street Journal. December 25, 2001.

King, D. 2002. Presentation to the NRC Animal Health Workshop. The British Battle Against Foot-and-Mouth Disease. January 2002.

Klinger, Terrie. 2002. Variability and uncertainty in crop-to-wild hybridization. Pp. 1-15 in Genetically Engineered Organisms, D.K. Letourneau and B.E. Burrows, eds. Boca Raton, FL: CRC Press

Kogan, M. 1998. Integrated pest management: Historical perspectives and contemporary developments. Annu. Rev. Entomol. 43:243-270.

Kohnen, A. 2000. Responding to the threat of agroterrorism: Specific recommendations for the United States Department of Agriculture. BCSIA Discussion Paper 2000-29, ESDP-2000-04. John F. Kennedy School of Government. Harvard University.

Landis, D.A., S.D. Wratten, and G.M. Gurr. 2000. Habitat management to conserve natural enemies of arthropod pests in agriculture. Annu. Rev. Entomol. 45:175-201.

Larkin, R.P., and D.R. Fravel. 1999. Mechanisms of action and dose-response relationships governing biological control of fusarium wilt of tomato by nonpathogenic *Fusarium* spp. Phytopathology 89(12):1152-1161.

Lautner, B. 2001. Presentation to Committee on Biological Threats to Agricultural Plants and Animals. August 15, 2001.

Layne, S.P., and T.J. Beugelsdijk. 1999. Laboratory firepower for infectious disease research. Nature Biotechnology 16:825-829.

Ligler, F.S., G.P. Anderson, P.T. Davidson, R.J. Foch, J.T. Ives, K.D. King, G. Page, D.A. Stenger, and J.P. Whelan. 1998. Remote Sensing Using an Airborne Biosensor. Environ Sci. Technol. 32(16):2461-2466.

Ligler, F.S., C.A. Rowe-Taitt, J.P. Golden, M.J. Feldstein, J.J. Cras, and K.E. Hoffman. 2000. Array biosensors for detection of biohazards. Biosensors and Bioelectronics. Special Issue 14:785-794.

Lorito, M., S.L. Woo, I.G. Fernandez, G. Colucci, G.E. Harman, J.A. Pintor-Toro, E. Filippone, S. Muccifora, C.B. Lawrence, A. Zoina, S. Tuzun, and F. Scala. 1998. Gene from mycoparasitic fungi as a source for improving plant resistance to fungal pathogens. Proc. Natl. Acad. Sci. 95:7860-7865.

Madden, L.V., and H. Scherm. 1999. Epidemiology and Risk Prediction. American Phytopathological Society Annual Meeting Symposium.

Madden, L.V., and F. van den Bosch. 2002. A population-dynamic approach to assess the threat of plant pathogens as biological weapons against annual crops. BioScience 52(1):65-74.

Malakoff, D. 2002. U.S. budget: Spending triples on terrorism R&D. Science 295:254.

Marshall, A., and J. Hodgson. 1998. DNA chips: An array of possibilities. Nature Biotechnology 16:27-31.

May, B.J., B. Zhang, L.L. Li, M.L. Pqustian, T.S. Whittam, and V. Kapur. 2001. Complete genomic sequence of *Pasteurella multocida,* Pm70. PNAS 98:3460-3465.

McLoughlin, M.P., W.R. Allmon, C.W. Anderson, A.A. Carlson, D.J. Dcicco, and N.H. Evancich. 1999. Johns Hopkins APL Technical Digest 20(3), July-September 1999:326-334.

Meadows, M.E. 1985. Eradication program in the southeastern United States. Pp. 8-11 in Symposium on eradication of the screwworm from the United States and Mexico. O.H. Graham, ed. Misc. Publ. 62. Entomol. Soc. Amer. College Park, MD.

Mitchell, C.J. 1995. Geographic Spread of Aedes albopictus and Potential for Involvement in Arbovirus Cycles in the Mediterranean Basin. J. Vector Ecol. 20:44-58.

Moseley, J.R. 2001. Role of the USDA in Homeland Security. Testimony before the House Agriculture Committee. November 15th, 2001.

Murch, R. 2001a. Presentation to the Committee on Biological Threats to Agricultural Plants and Animals. May 31, 2001.

Murch, R. 2001b. Forensic perspective on bioterrorism. In Firepower in the Lab. Automation in the Fight Against Infectious Diseases and Bioterrorism. S.P. Layne, T.J. Beuqelsdijk, and C.K.N. Patel, eds. Joseph Henry Press. Washington, DC: National Academy of Sciences:

NAS. 1987. Introduction of recombinant DNA-engineered organisms into the environment: Key issues. Washington, DC: National Academy Press.

NASDARF (National Association of State Departments of Agriculture Research Foundation). 2001. The Animal Health Safeguarding Review: Results and Recommendations, October 2001.

National Plant Board. 1999. Safeguarding American Plant Resources. Washington, DC: U.S. Department of Agriculture.

NCDA and CS (North Carolina Department of Agriculture and Consumer Services). 2002. North Carolina State Animal Response Team. [Online]. Available: http://www.ncsart.org/ [2002, February 28].

NRC. 1989. Field testing genetically modified organisms. Washington, DC: National Academy Press.

NRC. 1998. Ensuring Safe Food: From Production to Consumption. Washington, DC: National Academy Press

NRC. 1999. Chemical and Biological Terrorism: Research and Development to Improve Civilian Medical Response. Washington, DC: National Academy Press.

NRC. 2000. Genetically Modified Pest-protected Plants. Washington, DC: National Academy Press.

NRC. 2002. Predicting Invasions of Nonindigenous Plants and Plant Pests. Washington, DC: National Academy Press.

Neher, N. 1999. The Need for a coordinated repsonse to food terrorism: The Wisconsin Experience. Pp. 181 in Food and Agricultural Security, T.W. Frazier and D.C. Richardson, eds. Annals of the NY Academy of Sciences 894.

NIAID (National Institute for Allergy and Infectious Disease). 2001. Research on Medical Tools to Combat Bioterrorism. [Online]. Available: http://www.niaid.nih.gov/factsheets/btmedtools.htm.

NISC (National Invasive Species Council). 2001. Meeting the Invasive Species Challenge: National Management Plan. [Online]. Available: http://www.invasivespecies.gov/council/nmp.shtml.

Northrup, A. 2001. Presentation to Committee on Biological Threats to Agricultural Plants and Animals. August 15, 2001.

O'Donnell, K., H.C. Kistler, B.K. Tacke, and H.H. Casper. 2000. Gene geneologies reveal global phylogeographic structure and reproductive isolation among lineages of the fungus causing wheat scab. Proc. Nat. Acad. Sci. 97:7905-7910.

OECD (Organisation for Economic Cooperation and Development). 1993. Safety considerations for biotechnology: Scale-up of crop plants. Paris: OECD.

OIE (Office International des Epizooties). 2000. Manual of Standards, Diagnostic Tests, and Vaccines 2000: Rift Valley Fever. [Online]. Available: http://www.oie.int/eng/normes/mmanual/a_00029.htm.

OIE (Office International des Epizooties). 2002. OIE Web site. [Online]. Available: http://www.oie.int/eng/OIE/en_oie.htm.

Oldroyd, G.E.D., and B.J. Staskawicz. 1998. Genetically engineered broad-spectrum disease resistance in tomato. Proceedings of the National Academy of Sciences 95:10300-10305.

Park, S., T.A. Taton, and C. Mirkin. 2002. Array-based electrical detection of DNA with nanoparticle probes. Science 295:1503-1506.

PDD (Presidential Decision Directive). 1995. U.S. Policy on Counterterrorism. [Online]. Available: http://www.fas.org/irp/offdocs/pdd39.htm [1995, June 21].

Powers, L., and W. Ellis, Jr. 1998. Pathogenic microbe sensor technology. Presentation at the DARPA meeting on Biosurveillance: Providing Detection in the New Millenium. February 11. Laurel, MD: Johns Hopkins University Applied Physics Laboratory.

Rendeleman, C.M., and F.J. Spinelli. 1994. An economic assessment of the costs and benefits of African swine fever prevention. Animal Health Insight. 1994 Spring/Summer:18-27.

Richardson, R.H., and W.W. Averhoff. 1978. A cattleman's view of beef production with and without screwworms. Pp. 3-9 in The Screwworm Problem. Evolution of Resistance to Biological Control, R.H. Richardson, ed. Austin, TX: University Texas Press.

Rogers, P., S. Whitby, and M. Dando. 1999. Biological warfare against crops. Scientific American June: 70-75.

Sakai, A.K., F.W. Allendorf, J.S. Holt, D.M. Lodge, J. Molofsky, K.A. With, S. Baughman, R.J. Cabin, J.E. Cohen, N.C. Ellstrand, D.E. McCauley, P. O'Neill, I.M. Parker, J.N. Thompson, S.G. Weller. 2001. The population biology of invasive species. Annu. Rev. Ecol. Syst. 32:305-332.

Schubert, T.S., S.A. Rizvi, X. Sun, T.R. Gottwald, J.H. Graham, and W.N. Dixon. 2001. Meeting the challenge of eradicating citrus canker in Florida—again. Plant Disease 85:340-356.

Sequeira, R. 1999. Safeguarding production agriculture and natural ecosystems against biological terrorism. Pp. 48 in Food and Agricultural Security, T.W. Frazier and D.C. Richardson, eds. Annals of the NY Academy of Sciences 894.

Simpson, A.J., F.C. Reinach, P. Arruda, F.A. Abreu, M. Acencio, R. Alvarenga, L.M. Alves, J.E. Araya, G.S. Baia, C.S. Baptista, M.H. Barros, E.D. Bonaccorsi, S. Bordin, J.M. Bove, M.R. Briones, M.R. Bueno, A.A. Camargo, L.E. Camargo, D.M. Carraro, H. Carrer, N.B. Colauto, C. Colombo, F.F. Costa, M.C. Costa, C.M. Costa-Neto, L.L. Coutinho, M. Cristofani, E. Dias-Neto, C. Docena, H. El-Dorry, A.P. Facincani, A.J. Ferreira, V.C. Ferreira, J.A. Ferro, J.S. Fraga, S.C. Franca, M.C. Franco, M. Frohme, L.R. Furlan, M. Garnier, G.H. Goldman, M.H. Goldman, S.L. Gomes, A. Gruber, P.L. Ho, J.D. Hoheisel, M.L. Junqueira, E.L. Kemper, J.P. Kitajima, J.E. Krieger, E.E. Kuramae, F. Laigret, M.R. Lambais, L.C. Leite, E.G. Lemos, M.V. Lemos, S.A. Lopes, C.R. Lopes, J.A. Machado, M.A. Machado, A.M. Madeira, H.M. Madeira, C.L. Marino, M.V. Marques, E.A. Martins, E.M. Martins, A.Y. Matsukuma, C.F. Menck, E.C. Miracca, C.Y. Miyaki, C.B. Monteriro-Vitorello, D.H. Moon, M.A. Nagai, A.L. Nascimento, L.E. Netto, A. Nhani, Jr., F.G. Nobrega, L.R. Nunes, M.A. Oliveira, M.C. de Oliveira, R.C. de Oliveira, D.A. Palmieri, A. Paris, B.R. Peixoto, G.A. Pereira, H.A. Pereira, Jr., J.B. Pesquero, R.B. Quaggio, P.G. Roberto, V. Rodrigues, A.J. de M Rosa, V.E. de Rosa, Jr., R.G. de Sa, R.V. Santelli, H.E. Sawasaki, A.C. da Silva, A.M. da Silva, F.R. da Silva, W.A. da Silva, Jr., J.F. da Silveira, M.L. Silvestri, W.J. Siqueira, A.A. de Souza, A.P. de Souza, M.F. Terenzi, D. Truffi, S.M. Tsai, M.H. Tsuhako, H. Vallada, M.A. Van Sluys, S. Verjovski-Almeida, A.L. Vettore, M.A. Zago, M. Zatz, J. Meidanis, J.C. Setubal. 2000. The genome sequence of the plant pathogen *Xylella fastidiosa*. The *Xylella fastidiosa* Consortium of the Organization for Nucleotide Sequencing and Analysis. Nature 406(6792):151-157.

Smith, M.E., E.O. van Ravenswaay, and S.R. Thompson. 1988. Sales loss determination in food contamination incidents: An application to milk bans in Hawaii. American Journal of Agricultural Economics. 70:513-520.

Swartz, D.G., and I.E. Strand. 1981. Avoidance costs associated with imperfect information: The case of kepone. Land Economics 57:139-150.

Tauzin, W.J. 2002. Committee News Release. U.S. House of Representatives Committee on Energy and Commerce. [Online]. Available: http://energycommerce.house.gov/107/news/05222002_577print.htm.

Thaler, R. 1980. Toward a positive theory of consumer choice. Journal of Economic Behavior and Organization 1:39-60.

TIGR (The Institute for Genomic Research). 2002. Microbial Database: Completed, published microbial genomes. [online]. Available: http://www.tigr.org/tdb/mdb/mdbcomplete.html.

USAHA (United States Animal Health Association). 2002. National Animal Health Emergency Management System. [Online]. Available: http://www.usaha.org/NAHEMS/ [2002, February 6].

US Congress. 1884. Regulations for suppression of diseases; cooperation of States and Territories. Sec. 3, 23 Stat. 32. 21 USC 114.

US Congress. 1903. Regulations to prevent contagious disease. Sec. 2, 32 Stat. 792. 21 USC 111.

US Congress. 1962. Seizure, quarantine, and disposal of livestock or poultry to guard against introduction or dissemination of communicable diseases. Sec. 2, PL 87-518, 76 Stat. 129. 21 USC 134a(d).

US Congress. 2000. Plant Protection Act. Title IV of Public Law 106-224, 114 Stat.438, 7 U.S.C. 7701-772, enacted June 20, 2000.

US Congress. 2001. Animal Health Protection Act. Bill H.R. 2002 [still pending].

USDA. 1998. Agricultural Marketing Service, Transportation and Marketing. Transport of US Grains. A Modal Share Analysis. 1978-95. March 1998.

USDA. 2000a. USDA, Livestock Production Distribution and Income.

USDA. 2000b. 2000 Annual Program Performance Report. [Online]. Available: http://www.usda.gov/ocfo/ar2000/aprpdf/araphis.pdf [2002, February 6].

USDA. 2001a. Food and Agricultural Policy: Taking Stock for the New Century. [Online]. Available: http://www.usda.gov/news/pubs/farmpolicy01/fpindex.htm.

USDA. 2002. News Release, January 31, 2002. [Online]. Available: http://www.usda.gov/news/releases/2002/01/0026.htm.

USDA-APHIS. 1998a. Hog Cholera. [Online]. Available: www.aphis.usda.gov/oa/pubs/fshc.html.

USDA-APHIS. 1998b. Foreign Animal Disease Report, Summer 1998. [Online]. Available: http://www.aphis.usda.gov/oa/pubs/fadrep.pdf.

USDA-APHIS. 1999a. Nipah Virus, Malaysia, May 1999 Emerging Disease Notice. [Online]. Available: http://www.aphis.usda.gov/vs/ceah/cei/nipah.htm [2002, February 6].

USDA-APHIS. 1999b. Inside APHIS Awards: Veterinary Services Awarded Prestigious W. Edward Deming Award. [Online]. Available: http://www.aphis.usda.gov/lpa/inside_aphis/awards2.html [2002, February 28].

USDA-APHIS. 2001a. Operation Labor Day. [Online]. Available: http://www.aphis.usda.gov/lpa/inside_aphis/featureMEGHAN.html.

USDA-APHIS. 2001b. Policy and Program Development Budget Tables.

USDA-APHIS. 2001c. Highly Pathogenic Avian Influenza. [Online]. Available: http://www.aphis.usda.gov/oa/pubs/avianflu.html [2002, February 6].

USDA-APHIS. 2001d. Karnal Bunt: A Fungal Disease of Wheat. [Online]. Available: http://www/aphis.usda.gov/oa/pubs/karnel.html [2002, February 6].

USDA-APHIS. 2001e. Veterinary Services Emergency Programs Incident Information Page. [Online]. Available: http://www.aphis.usda.gov/vs/ep/incident-info.html [2002, February 6].

USDA-APHIS. 2001f. APHIS Regulated Pest List. [Online]. Available: http://www.aphis.usda.gov/ppq/regpestlist/ [2002, February 6].

USDA-APHIS. 2001g. The Regulation of Transgenic Arthropods and Other Transgenic Invertebrates. [Online]. Available: http://www.aphis.usda.gov:80/bbep/bp/arthropod/#tgenadoc.

USDA-APHIS. 2001h. Foot and Mouth Disease Prevention Information. [Online]. Available: http://www.aphis.usda.gov/oa/fmd/owninfo.html.

USDA-APHIS. 2001i. Accreditation Standards for Laboratory Seed Health Testing and Seed Crop Phytosanitary Inspection. 66 Fed. Reg. 37397.

USDA-APHIS. 2001j. Restrictions on Importation of Restricted Articles. 7 CFR 319.76.

USDA-APHIS-PPQ. 1991a. Exotic Pest Detection Manual 1991.

USDA-APHIS-PPQ. 1991b. National Exotic Fruit Fly Trapping Protocol 1991.

USDA-APHIS-PPQ. 2000. Poster presentation on plant pest and pathogen interception at annual meeting of Entomological Society of America, December 3-6, 2000.

USDA-APHIS-PPQ. 2001a. Moratorium on issuing permits for importation of certain organisms. [Online]. Available: http://www.aphis.usda.ppq.gov/ppq/moratorium.pdf.

USDA-APHIS-PPQ. 2001b. Port Information Network. Accessed by APHIS-PPQ December 2001.

USDA-APHIS-PPQ. 2002. Emergency Programs Manual. [Online]. Available: http://www.aphis.usda.gov/ppq/emergencyprograms/manual.pdf [2002, February 28].

USDA-APHIS-VS. 2000. Veterinary Services Factsheet: Questions and Answers About Tripartite Exercise 2000. [Online]. Available: http://www.aphis.usda.gov/oa/tripart/qatripar.pdf [2002, February 11].

USDA-APHIS-VS. 2001. DRAFT Federal Emergency Response Plan for an Outbreak of Foot-and Mouth Disease or Other Highly Contagious Diseases. Washington, DC.

USDA-ARS. 2002. Agricultural Research Service Homepage. [Online]. Available: http://www.ars.usda.gov.

USDA-ERS. 2001a. Receipts of 10 leading states in 25 top commodities, 1970-2000. [Online] Available: http://www.ers.usda.gov/data/farmincome/firkdmu.htm [2002, March 13].

USDA-ERS. 2001b. U.S. Agricultural Trade. [Online]. Available: http://www.ers.usda.gov/briefing/AgTrade/usagriculturaltrade.htm.

USDA-ERS. 2001c. Wheat Situation and Outlook Yearbook. Washington, DC: Economic Research Service, U.S. Department of Agriculture.

USDA-ERS. 2002a. Agricultural Outlook, January-February. [Online]. Available: http://www.ers.usda.gov/publications/AgOutlook/Jan2002/.

USDA-ERS. 2002b. U.S. Agricultural Trade Update,. [Online]. Available: http://www.ers.usda.gov/publications/so/view.asp?f=trade/fau-bb/ [2002, January 22].

USDA-ERS. 2002c. Food Marketing and Price Spreads: Farm-to-Retail Price Spreads for Individual Food Items. [Online]. Available: http://www.ers.usda.gov/briefing/foodpricespreads/spreads/table1a.htm [2002, February 25].

USDA-ERS. 2002d. U.S. Agricultural Update. [Online]. Available: http://www.ers.usda.gov/publications/so/view.asp?f=trade/fau-bb/ [2002, February 25].

van den Bosch, F., J.A.J. Metz, and J.C. Zadoks. 1999. Pandemics of focal plant disease, a model. Phytopathology 89:495-505.

Van Ravenswaay, E.O. and J.P. Hoehn. 1991. The impact of health risk information on food demand: A case study of alar and apples. Pp. 155-174 in Economics of Food Safety, J.A. Caswell, ed. New York: Elsevier.

Veneman, A. 2001a. USDA's Role in Combating Terrorism. Testimony before Senate Committee on Appropriations Commerce, Justice, State, and the Judiciary Subcommittee. May 9, 2001.

Veneman, A. 2001b. Emergency Animal Disease Preparedness Activities. USDA news release. October 11, 2001.

Veneman, A. 2001c. Strengthening Biosecurity Measures. USDA news release. October 19, 2001.

Viscusi, W.K. 1997. Alarmist decisions with divergent risk information. Economic Journal 107:1657-1670.

Walt, D., and D. Franz. 2000. Biological warfare detection. Analytical Chemistry 72(230):738A-746A.

Weiss, R. 2002. Variance in Anthrax Strains Could Crack Case. Washington Post, February 18, 2002.

Wilson, T.M., L. Logan-Henfrey, R. Weller, and B. Kellman. 2000. Agroterrorism, biological crimes, and biological warfare targeting animal agriculture. Pp 23-57 in Emerging Diseases of Animals, C. Brown and C. Bolin, eds. Washington, DC: ASM Press.

Wendels, C.E. 2000. Economic and social impacts of Fusarium head blight: Changing farms and rural communities in the northern Great Plains. Phytopathology 90:17-21.

Wood, D.W., J.C. Setubal, R. Kaul, D.E. Monks, J.P. Kitajima, V.K. Okura, Y. Zhou, L. Chen, G.E. Wood, N.F. Almeida, Jr., L. Woo, Y. Chen, I.T. Paulsen, J.A. Eisen, P.D. Karp, D. Bovee, Sr., P. Chapman, J. Clendenning, G. Deatherage, W. Gillet, C. Grant, T. Kutyavin, R. Levy, M.J. Li, E. McClelland, A. Palmieri, C. Raymond, G. Rouse, C. Saenphimmachak, Z. Wu, P. Romero, D. Gordon, S. Zhang, H. Yoo, Y. Tao, P. Biddle, M. Jung, W. Krespan, M. Perry, B. Gordon-Kamm, L. Liao, S. Kim, C. Hendrick, Z.Y. Zhao, M. Dolan, F. Chumley, S.V. Tingey, J.F. Tomb, M.P. Gordon, M.V. Olson, and E.W. Nester. 2001. The genome of the natural genetic engineer *Agrobacterium tumefaciens* C58. Science 294:2317-2323.

Appendixes

Appendix A

National Security Implications of Advances in Biotechnology: Threats to Plants and Animals

NATIONAL RESEARCH COUNCIL
COMMISSION ON LIFE SCIENCES
NAS Building
2101 Constitution Ave., NW
Washington, DC
AUGUST 5, 1999

Participants[1]

Dr. Linda Abbott
Ecologist
Office of Risk Assessment and Cost Benefit Analysis
US Department of Agriculture

Dr. John Bailar III
Chairman
Department of Health Studies
University of Chicago

[1]Note: Affiliations of the participants as of 8/5/99. Current affiliations may have changed.

Dr. Paul Berg
Chairman, Beckman Center
Stanford University School of Medicine

Dr. Roger Breeze
Director, South Atlantic Area
Agricultural Research Service
US Department of Agriculture

Dr. Neville P. Clarke
Professor
College of Veterinary Medicine
Texas A&M University

Dr. R. James Cook
Endowed Chair in Wheat Research
Department of Plant Pathology, Crops, & Soils
Washington State University

Dr. Richard Dunkle
Deputy Administrator, PPQ
Animal and Plant Health Inspection Service
US Department of Agriculture

Dr. Gerald Epstein
Senior Policy Analyst
Office of Science Technology Policy
The White House

Dr. Barry Erlick
Chief Scientist, Special Interagency Programs
Agricultural Research Service
US Department of Agriculture

Dr. David Franz
Vice President
Chemical and Biological Defense Division
Southern Research Institute

Dr. Thomas Frazier
President
GenCon

Dr. Sheldon K. Friedlander
Professor
Department of Chemical Engineering
University of California

Dr. Thomas Gingeras
Vice President, Biological Sciences
Affymetrix

Ms. Elisa Harris
Director, Non-Proliferation and Export Control
National Security Council

Dr. David Huxsoll
Professor
School of Veterinary Medicine
Louisiana State University

Dr. Shaun Jones
Program Manager
Defense Advanced Research Projects Agency

Dr. Paul Kaminski
Chair and C.E.O.
Technovation, Inc.

Dr. Kenneth Keller
Professor
Hubert H. Humphrey Institute of Public Affairs
University of Minnesota

Dr. Lonnie J. King
Dean
College of Veterinary Medicine
Michigan State University

Dr. Todd R. Klaenhammer
Professor
Department of Food Science
North Carolina State University

Dr. Harvey Ko
Chief Scientist and Program Manager for Advanced Technology
Applied Physics Laboratory
The Johns Hopkins University

Dr. Catherine Laughlin
Chief, Virology Branch
Division of Microbiology and Infectious Diseases, NIAID, NIH

Dr. Joshua Lederberg
Professor Emeritus
Laboratory of Molecular Genetics and Informatics
The Rockefeller University

Dr. Peter B. Merkle
Technical Advisor
Advanced Systems & Concepts Office
Defense Threat Reduction Agency

Dr. Harley Moon
F.K. Ramsey Chair of Veterinary Medicine
Veterinary Medical Research Institute
Iowa State University

Maj. Gen. Frank Moore
Deputy Director
Defense Threat Reduction Agency

Dr. Stephen S. Morse
Program Manager
Defense Advanced Research Projects Agency

Dr. Randall Murch
Deputy Assistant Director
Forensic Analysis Branch
Laboratory Division
Federal Bureau of Investigation

Dr. John Podlesny
Operations Research Analyst
HMRU-LAB, ERF
FBI Academy

Dr. Craig Reed
Administrator
Animal and Plant Health Inspection Service
US Department of Agriculture

Dr. Caird E. Rexroad, Jr.
Associate Deputy Administrator, Animal Production, Product Value and Safety
USDA, National Program Staff

Dr. Ronald Sederoff
Professor
Department of Forestry
North Carolina State University

Dr. Michael L. Shuler
Professor
School of Chemical Engineering
Cornell University

Dr. Norman Steele
BARC-East
Agricultural Research Service
US Department of Agriculture

Mr. Mark Urlaub
Intelligence Analyst
Department of Defense

Dr. Catherine E. Woteki
Undersecretary of Food Safety
US Department of Agriculture

Board on Biology Staff

Dr. Ralph Dell
Acting Director

Dr. Jennifer Kuzma
Project Director

Ms. Kathleen Beil
Acting Administrative Assistant

Ms. Nina Mani
NRC Summer Intern

Board on Agriculture and Natural Resources Staff

Ms. Charlotte Kirk Baer
Acting Director

COMMISSION ON LIFE SCIENCES AND
supported by the
COMMISSION ON PHYSICAL SCIENCES, MATHEMATICS, AND APPLICATIONS
COMMITTEE ON INTERNATIONAL SECURITY AND ARMS CONTROL

National Security Implications of Advances in Biotechnology: Threats to Plants and Animals

A Planning Meeting
August 5, 1999
Washington, DC

Summary

A one-day planning meeting will be hosted by the National Research Council (NRC) to explore the linkages between advances in biotechnology and the use of and defenses against biological weapons, especially those that would target plants or animals. The meeting will determine whether, the NRC using industrial and academic researchers, can play a further role in assisting the federal government in preventing and responding to such threats. The planning meeting will be organized by the Commission on Life Sciences, with the assistance of the Committee on International Security and Arms Control, and the Commission on Physical Sciences, Mathematics, and Applications. The product of the meeting will be an internal document advising NRC of the group's recommendations concerning whether a further activity is appropriate and, if so, what form it should take.

National Security Implications of Advances in Biotechnology

National Research Council Steering Committee

John Bailar III, Ph.D., MD
Professor
Department of Health Studies
University of Chicago

Paul Berg, Ph.D.
Chairman
Beckman Center
Stanford University School of Medicine

David Goeddel, Ph.D.
Chief Executive Officer
Tularik, Inc.

Paul Kaminski, Ph.D.
Chair and C.E.O.
Technovation, Inc.

Kenneth Keller, Ph.D.
Professor of Science and Technology Policy
Hubert H. Humphrey Institute of Public Affairs
University of Minnesota

Harley Moon, Ph.D., D.V.M.
F.K. Ramsey Chair, Veterinary Medicine
Veterinary Medical Research Institute
Iowa State University

Ronald Sederoff, Ph.D.
Edwin F. Conger Professor of Forestry
Department of Forestry
North Carolina State University

National Security Implications of Advances in Biotechnology: Threats to Plants and Animals

A Planning Meeting
August 5, 1999
NAS Building
Washington, DC

Welcome from Workshop Co-Chairs and Introduction of Participants

8:30-9:00 am *Co-chairs,*
Harley Moon, Iowa State University
Paul Berg, Stanford University

I. Overview of Key Concerns and Programs from DOD and USDA
Chair, Harley Moon

9:00-10:00 DOD concerns and programs for plant and animal threats
Major General Frank Moore, Deputy Director, Defense Threat Reduction Agency
Unclassified Briefing: Mark Urlaub, Intelligence Analyst

10:00-10:15 **BREAK**

10:15-11:15 USDA concerns and programs for plant and animal threats
Roger Breeze, Director, South Atlantic Area, Agricultural Research Service
Catherine Woteki, Undersecretary of Food Safety
Craig Reed, Administrator, Animal and Plant Health Inspection Service

Opening Remarks for the Discussion Sessions

11:15-11:30 *Joshua Lederberg, The Rockefeller University*

II. Discussion of Possible Scenarios and Impacts
Chair, Harley Moon

11:30-12:30 pm *What are possible scenarios for attacks on plants or animals?*
What are potential impacts of these scenarios?

What are the differences between perceived impacts and probable impacts?
What government programs are in place for detection, prevention, response, and recovery?
What are agency needs for detection, prevention, response, and recovery?

12:30-1:30 pm **LUNCH**

III. Discussion of Relevant Research in Academe and Industry
Chair, Paul Berg

1:30-3:00 *What do we know about the science of the creation, distribution, detection, and prevention of biological threats to plants and animals?*
What methods are available for effective detection, prevention, and recovery?
What do we need to know for effective detection, prevention, and recovery?
What are priorities for research and development in these areas?
What are future strategies for research and development in these areas?

3:00-3:30 **BREAK (submission of ideas for session IV)**

IV. Discussion of Priority Areas and Possible NRC Activities
Chair, Paul Berg

3:30-4:30 *How can academic and industry scientists contribute to national security concerning threats to plants and animals?*
Does the NRC have a role to play in fostering these contributions?
If so, via what mechanism?
(e.g. hosting a larger workshop?, consensus study?)

What is the role of the public and private sectors in addressing biotechnology as a concern and as a solution to bioterrorism?

Does the NRC have a role to play in providing advice to the agencies, industry and/or academe on prevention and detection of, response to, and recovery from biological threats to plants and animals?
If so via what mechanism?
(e.g. hosting a larger workshop?, consensus study?)

Appendix B

Public Meetings of the Committee on Biological Threats to Agricultural Plants and Animals

OPEN SESSION OF THE FIRST COMMITTEE MEETING
NATIONAL RESEARCH COUNCIL
BOARD ON AGRICULTURE AND NATURAL RESOURCES

National Academy of Sciences
2101 Constitution Ave NW
Washington, DC
May 31, 2001

Agenda

1:00-2:00 pm — An Overview of Important Study Issues and USDA's Interest in this Study
Roger Breeze, Associate Administrator, USDA-ARS

2:00-3:00 pm — An Overview of the Defense and Intelligence Issues Related to Agricultural Bioterrorism
Randy Murch, Department of Defense, Defense Threat Reduction Agency

ROUNDTABLE DISCUSSION WITH THE COMMITTEE ON BIOLOGICAL THREATS TO AGRICULTURAL PLANTS AND ANIMALS
NATIONAL RESEARCH COUNCIL
BOARD ON AGRICULTURE AND NATURAL RESOURCES
AUGUST 14, 2001

National Academy of Sciences
2101 Constitution Ave NW
Washington, DC

Agenda

Roundtable purpose For the committee to gather information from outside experts about the U.S. defense system for agriculture and its use of science.

Roundtable format 6-8 participants will be invited to join the committee at the table for each session. Participants will be asked to give five-minute opening remarks (time strictly limited). Participants will be asked to engage in discussion with the committee.

Session I—9:00 to 11:00 a.m.

Overview of U.S. defense system for agriculture
- Tom Frazier, GenCon
- Terrance Wilson, USDA liaison, Armed Forces Medical Intelligence Center, Ft. Detrick

Investigations of threats to agriculture
- Joyce Fleischman, USDA Office of the Inspector General

Experiences with and capabilities to prevent or manage plant and animal threats at the state and local levels
- Tracey McNamara, Bronx Zoo
- Nicholas Neher, Wisconsin Department of Agriculture
- Bob Hillman, State Veterinarian of Idaho
- Julie Beale, University of Kentucky—Extension, Plant Disease Diagnostic Laboratory
- John Sullivan, Los Angeles County Sheriff's Emergency Operations Bureau

Session II—12:30 to 2:30 p.m.

Overview of U.S. federal system for agricultural defense
Alfonso Torres, USDA-APHIS, Veterinary Services
Gordon Gordh, USDA-APHIS, Plant Protection and Quarantine
Walter Hill, USDA, Food Safety and Inspection Service
Clifford Oliver, USDA Office of Crisis Planning and Management
Dan McChesney, FDA Center for Veterinary Medicine
Tom Kinnally, National Domestic Preparedness Office
Richard Meyer, CDC
Robert McCreight, State Department

Session III—3:30 to 5:30 p.m.

Intelligence and defense capabilities
Peter Probst, Institute for the Study of Terrorism
Scientific perspectives on U.S. capabilities
David Huxsoll, USDA-ARS Plum Island
Norman Schaad, Ft. Detrick, USDA-ARS
Richard Orr, National Invasive Species Council
Industry: roles, concerns, & science
Beth Lautner, National Pork Board
Allen Northrup, Cepheid Inc.
Vicki Hall, Ft. Dodge Animal Health Laboratories
Mark Condon, American Seed Trade Association

ROUNDTABLE DISCUSSION WITH THE COMMITTEE ON BIOLOGICAL THREATS TO AGRICULTURAL PLANTS AND ANIMALS NATIONAL RESEARCH COUNCIL BOARD ON AGRICULTURE AND NATURAL RESOURCES

Discussion Questions[1]

The U.S. Defense System—

–How are you and your organization involved (or how do you plan to be involved) in deterring, preventing, detecting, thwarting, responding to, and recovering from biological threats to plants and animals (whether natural or intentional)?

–What interagency/inter-organizational efforts are underway to develop a coordinated defense plan for agriculture?

–Are there any historical examples of biological threats to plants or animals (natural or intentional) that might be useful for the committee to analyze? If so, how did the U.S. defense system coordinate/operate and use science/technology to deter, prevent, thwart, detect, respond, and recover?

–How does the U.S. defense system plan to respond to plant or animal disease? (e.g. FMD responses such as culling and killing animals, vaccination, etc.)

–What is the role of the private sector? What efforts are underway in this sector?

–What are the overall strengths and weaknesses of the U.S. defense system for agriculture?

The Use of Science—

–What are the current scientific capabilities of the U.S. defense system?
- –recognition of an incident
- –traceback to origin
- –detection of agricultural BW agents

[1]Note—each participant might not be able to address all the questions. Please choose the ones closest to your expertise/experiences. Thank you.

–prophylactic measures
  –decontamination procedures
  –treatments/vaccines
  –containment
  –etc., etc.

–Does the U.S. defense system have access to and/or is it using current (best available) science and technology in order to deter, prevent, thwart, detect, respond and recover, including both public and private capabilities? If not, what are the barriers to implementation and what needs to be done in order to ensure it?

–What data, scientific research or technologic developments are needed to improve defense for biological threats to plants and animals?

–What improvements in educating and/or training personnel and the public are needed to improve the defense system?

OPEN SESSION OF THE THIRD COMMITTEE MEETING
NATIONAL RESEARCH COUNCIL
BOARD ON AGRICULTURE AND NATURAL RESOURCES
NOVEMBER 15, 2001

National Academy of Sciences
2001 Wisconsin Ave NW
Washington, DC

Agenda

1:00-1:30 pm	Social and Psychological Aspects of Threats to Agriculture *Steven M. Becker, University of Alabama at Birmingham*
1:30-2:00 pm	Global Surveillance Laboratory against Bioterrorism: FMD Laboratory Surveillance *Scott P. Layne, University of California, Los Angeles*

Appendix C

Acronym List

ARC	American Red Cross
ASTA	American Seed Trade Association
BSE	bovine spongiform encephalopathy ("Mad Cow Disease")
CAPS	Federal/State Cooperative Agricultural Pest Survey (under APHIS-PPQ)
CCEP	Citrus Canker Eradication Program
CDC	Centers for Disease Control and Prevention
CJD	Creutzfeld-Jakob disease
DNA	deoxyribonucleic acid
DOC	Department of Commerce
DOD	Department of Defense
DOE	Department of Energy
DOI	Department of Interior
DOJ	Department of Justice
DOL	Department of Labor
DOT	Department of Transportation
ELISA	enzyme-linked immunosorbent assay
EMS	emergency management system
EPA	Environmental Protection Agency
ERT	emergency response team
ESA	Entomological Society of America

FADDs	foreign animal disease diagnosticians
FADs	foreign animal diseases
FAO	Food and Agriculture Organization (under UN)
FBI	Federal Bureau of Investigation
FCA	Farm Credit Administration
FDA	Food and Drug Administration
FEMA	Federal Emergency Management Agency
FIS	Federal Inspection Service
FMD	foot-and-mouth disease
FMDV	foot-and-mouth disease virus
GMO	genetically-modified organism
GSA	General Services Administration
HHS	Department of Health of Human Services
INS	Immigration and Naturalization Service
IPPC	International Plant Protection Convention
NAPIS	National Agricultural Plant Information System (under USDA-APHIS)
NAPPO	North American Plant Protection Organization (under IPPC)
NASDA	National Association of State Departments of Agriculture
NCS	National Communications System
NEDSS	National Electronic Disease Surveillance System
NIC	National Intelligence Council
NPAG	New Pest Advisory Group (under APHIS)
NRC	National Research Council
NSHS	National Seed Health System (under APHIS)
OIE	Office International des Epizooties
PCR	polymerase chain reaction
PNW	Pacific Northwest
POE	port of entry
PPA	Plant Protection Act
PRP	protease resistant protein
PSbMV	pea seedborne mosaic virus
READEO	Regional Emergency Animal Disease Eradication Organization
RNA	ribonucleic acid

RRT	Regional Rapid Response Teams
RVF	Rift Valley Fever
SAW	surface acoustic wave
SBA	Small Business Administration
SDOA	State Departments of Agriculture
SIRM	sterile insect release method
SIT	sterile insect technique
SLHD	state and local health departments
TCK	*Tilletia controversa* Kuhn
TSE	transmissible spongiform encephalopathies
UN	United Nations
USAMRMC	United States Army Medical Research and Material Command
USDA	United States Department of Agriculture
USDA-APHIS	USDA Animal and Plant Health Inspection Service
USDA-APHIS-PPQ	USDA APHIS Plant Protection and Quarantine
USDA-APHIS-VA	USDA APHIS Veterinary Services
USDA-ARS	USDA Agricultural Research Service
USDA-ERS	USDA Economic Research Service
USDA-FSA	Farm Services Agency
USDA-FSIS	USDA Food Safety and Inspection Service
USDA-OIG	USDA Office of Inspector General
vCJD	variant Creutzfeld-Jakob disease
VVND	velogenic viscerotropic Newcastle disease

Appendix D

United States Department of Agriculture Homeland Security Council

DEPARTMENT OF AGRICULTURE
OFFICE OF THE SECRETARY
WASHINGTON, D.C. 20250

January 10, 2002

Mr. Mark Holman
Deputy Assistant to the President for
Homeland Security
Office of Homeland Security
The White House
Washington, DC 20502

Dear Mark:

One of Secretary Veneman's priorities is to ensure that USDA's multiple mission areas are well coordinated on issues of common interest with other Cabinet agencies. This is particularly true with regard to the Department's efforts to meet the needs and directives of the Office of Homeland Security (OHS).

Following the structure developed for OHS by Governor Ridge, we established the U.S. Department of Agriculture Homeland Security Council (USDA/HSC) last November. This internal council, chaired by Deputy Secretary Jim Moseley, has the mandate to centralize and coordinate all the USDA homeland security activities under a single departmental authority at a senior leadership level to enhance rapid response plans, improve communications, and ensure best use of all available resources.

To assist Governor Ridge, you and your staff in identifying USDA contacts/sources for meetings, information and communications requests, Departmental updates, etc., the Secretary has designated the following initial points-of-contact on all Homeland Security issues to ensure that we are meeting OHS needs:

Deputy Secretary Jim Moseley (Shelia Trollinger, Confidential Assistant & Scheduler)
(phone) 202-720-6158; (fax) 202-720-5437; (email) Shelia.Trollinger@usda.gov

Deborah Atwood, Special Assistant to the Deputy Secretary
(phone) 202-720-6158; (fax) 202-720-5437; (email) Deb.Atwood@usda.gov

Dr. Curt Mann, Special Assistant to the Deputy Secretary
(phone) 202-720-3804; (fax) 202-720-5437; (email) Curt.Mann@usda.gov

Dale Moore, Chief of Staff, USDA
(phone) 202-720-7971; (fax) 202-690-2119; (email) dwm@usda.gov

AN EQUAL OPPORTUNITY EMPLOYER

-2-

This direct operational interaction through the Deputy Secretary and the USDA/HSC senior staff will not only shorten the lines of communication, but will increase USDA's effectiveness in working with OHS to carry out our responsibilities, responses, and actions pertaining to homeland security in the most expeditious and effective manner possible. If you have any questions or suggestions for further improvements, please do not hesitate to contact me.

Homeland Security Council senior staff contact information:

Deborah Atwood
Special Assistant to the Deputy Secretary (202-720-6158)

Curt Mann
Special Assistant to the Secretary (202-720-3804)

Jeremy Stump
Confidential Assistant to the Chief of Staff (202-720-3275)

Barry Erlick
Chief Scientist, Special Research Programs (202-690-1476)

Keith Collins
Chief Economist (202-720-4164)

Sincerely,

Dale W. Moore
Chief of Staff

Enclosures

DEPARTMENT OF AGRICULTURE
OFFICE OF THE DEPUTY SECRETARY
WASHINGTON, D.C. 20250

TO: SECRETARY VENEMAN
SUBCABINET OFFICIALS
AGENCY HEADS

FROM: JIM MOSELEY
DEPUTY SECRETARY

DATE: NOVEMBER 15, 2001

SUBJECT: HOMELAND SECURITY COUNCIL

Attached you will find a copy of our USDA Homeland Security Council. Please hand out to anyone you deem appropriate.

Attachment

USDA Homeland Security

USDA has created a structure or the management of its homeland security responsibilities. This replaces structures established by the previous Administration for the management of anti-terrorism activities.

The structure incorporates the key USDA responsibilities identified in the Department's report to the Office of Homeland Security. These responsibilities are:

1. Protection of Borders and Agricultural Production
2. Food Safety
3. Protecting and Enhancing Research and Laboratory Facilities
4. Protecting USDA Staff and Other Infrastructure
5. Securing Information Technology Resources
6. Law Enforcement Activities and Audits
7. Emergency Preparedness

The structure of a USDA Homeland Security Council and three subcouncils provides coordination between mission areas and agencies as well as an adequate flow of information on day to day events to the Secretary and other key decision makers.

Each major agency will be asked to identify one individual with overall responsibility for homeland security issues, including working with counterparts from other agencies and to make institutional arrangements to assure that homeland security activities are coordinated within the agency

USDA Homeland Security Council

While the entire homeland security effort will be overseen by the Secretary, the USDA Homeland Security Council will be chaired by the Deputy Secretary in his capacity as the Department's Chief Operating Officer. The Council will include all Under and Assistant Secretaries, the Inspector General, select staff office directors, and the communications director. Staff support for this Council will be provided by a Special Assistant for Homeland Security to be appointed by the Secretary or the Deputy Secretary.

The Council is responsible for overall USDA homeland security policy, coordination of department-wide homeland security issues, approval of budgets or other major commitments; appointment of USDA representation to inter-agency or other external groups; tracking USDA progress on high priority homeland security objectives; resolving issues that cannot be resolved at a lower level; and external communications.

The Council will ensure that information, research, and resources are shared and activities coordinated with other Federal agencies as appropriate.

Subcouncils

Each of the three subcouncils will be chaired or co-chaired by USDA sub-cabinet officers. The subcouncils will have responsibility for necessary coordination among mission areas and agencies to carry out the particular activity, recommend budgets or other major commitments to the Homeland Security Council, recommend USDA representation to inter-agency or other external groups, and provide reports to the Homeland Security Council concerning the status of key objectives and issues which need to be resolved. In addition, all subcouncils are responsible for coordination with other related Federal and State Agencies.

Protection of the Food Supply and Agricultural Production Subcouncil

This subcouncil will be co-chaired by the Under Secretaries for Food Safety and Marketing and Regulatory Programs. This subcouncil is responsible for issues dealing with: food production, processing, storage, and distribution; threats against the agricultural sector and rapid response to such threats; border surveillance and protection to prevent introduction of plant and animal pests and diseases; food safety activities concerning meat, poultry, and egg inspection, laboratory support, research, education, etc., and outbreaks of foodborne illness.

Protecting USDA Facilities and Other Infrastructure

This subcouncil is co-chaired by the Under Secretaries for Natural Resources and Environment and Research, Education and Economics. It is responsible for issues dealing with security of all USDA facilities and equipment including the security of laboratory and technical facilities; biohazards; pathogens; collections and information; and scientists security; and National Forest System lands and related infrastructure. In addition, this subcouncil is responsible for all issues relating to the security of information technology resources of USDA.

Protecting USDA Staff/Emergency Preparedness Subcouncil

This subcouncil is chaired by the Assistant Secretary for Administration is responsible for issues dealing with the safety of all USDA staff; the coordination of the Continuity of Operations Plan (COOP); Continuity of government (COG); Occupant Emergency Planning (OEP); and Federal inter-agency emergency coordination. In addition, this subcouncil is responsible for emergency communications within USDA.

USDA HOMELAND SECURITY STRUCTURE

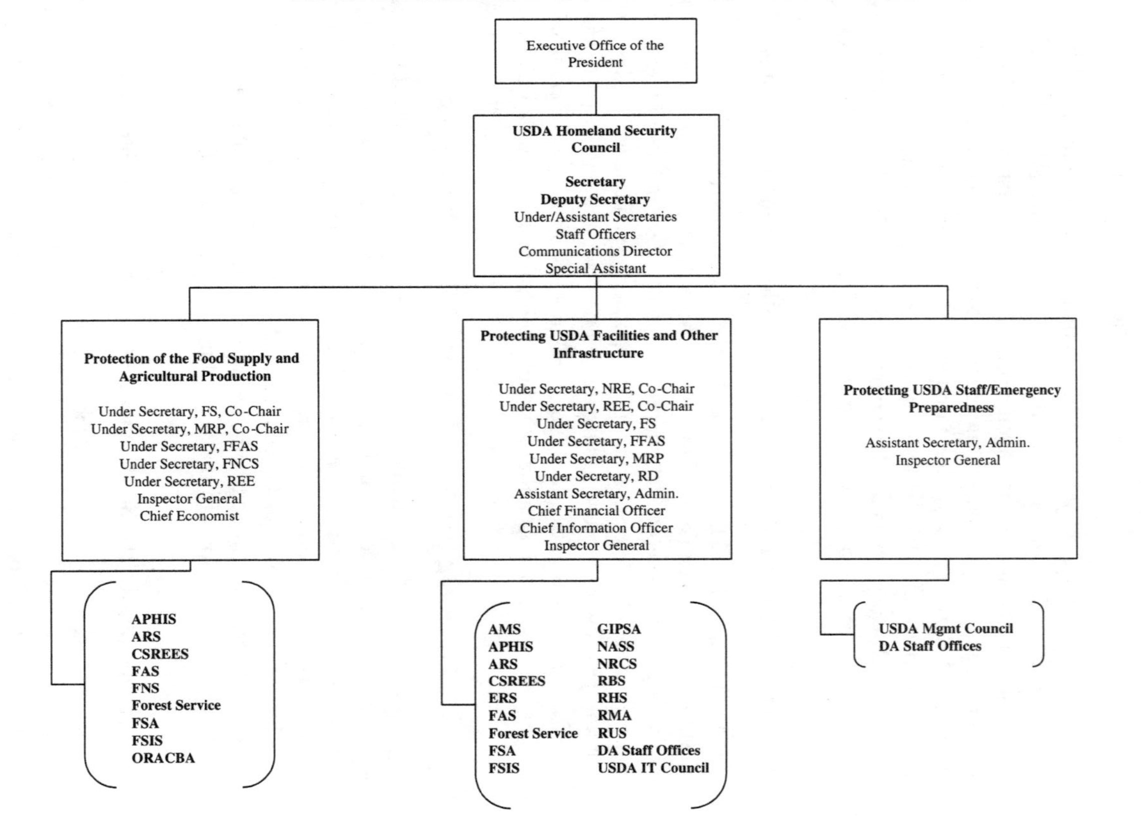

United States Department of Agriculture • Office of Communications • 1400 Independence Avenue, SW
Washington, DC 20250-1300 • Voice: (202) 720-4623 • Email: oc.news@usda.gov • Web: http://www.usda.

of

James R. Moseley Deputy Secretary, U.S. Department of Agriculture
Before the House Agriculture Committee
On Thursday, November 15, 2001

"Mr. Chairman and members of the Committee, it is an honor for me to appear before you today to discuss the important role played by the Department of Agriculture (USDA) in homeland security. I appreciate your initiative in calling this hearing because the Department's actions in support of homeland security are very important to farmers, consumers, and all of the other constituents of our programs.

"As you know, the President has taken decisive action to protect our homeland security in light of the events of September 11th. Executive Order 13228 established the Office of Homeland Security and the Homeland Security Council. The Office of Homeland Security is headed by former Pennsylvania Governor Tom Ridge. The Secretary and I have met with Governor Ridge and continue to provide his team with counsel and information about USDA's role in homeland security. The Secretary is a member of the Homeland Security Council, which is chaired by the President. The Secretary recently attended the first meeting of the Homeland Security Council and we at USDA are actively working to ensure the protection of our food supply. There is also a Working Group of Deputy Secretaries that has been established to support the Homeland Security Council, and I am a member of that Working Group. In short, the President has seen to it that we have the necessary institutions in place to assure the coordination and information flow that we will all need to carry out our homeland security responsibilities. In addition, he has requested $45.2 million in supplemental funding to further secure USDA facilities and programs.

"Mr. Chairman, USDA has a long history of assuring that the nation's supply of meat, poultry and egg products including production, processing, storage and distribution of foods, is safe and wholesome. My full testimony outlines the many areas of responsibilities of USDA in our food system.

"For instance, to date, we have prevented such devastating animal diseases as Foot and Mouth and BSE from entering this country. This has come as a result of a very dedicated team of animal health and plant health experts composed of federal, state and private efforts dedicated to maintaining our nations agricultural health. In fact, we implemented added security measures at the beginning of the year to prevent the spread of these diseases to the U.S. We added additional veterinarians and dog teams at ports of entry. We also increased the number of inspectors.

"Since September 11, we have worked in partnership with the Office of Homeland Security and the National Security Council as well as other Departments to set us on a course for long-term success. We have secured our facilities and inventoried our biological assets, with special emphasis on our labs across the country.

"We have sought input from a variety of interests to ensure we are addressing everything we need to. For instance, we continue to meet with industry, state officials, academia and others for input into the total response system.

"And, most importantly, we are communicating this information to people in face-to-face meetings, through the media and through our website.
My testimony today outlines many of the details in these areas. But I think it is important to also focus on our long-term plans and actions to prevent any threat that may occur.

"Our goal is to test our prevention and response systems across the board. To do this, we have organized an internal USDA Homeland Security Council chaired by myself with members from all of our program areas to ensure coordination across the department. In addition we are assessing our research needs to allow us to employ the latest technology to help in our efforts. And, we are formalizing a communication process to disseminate information about the products we regulate throughout the food chain. This will maintain confidence that we are doing everything possible to secure the products under our jurisdiction.

"Mr. Chairman, please let me now expand in certain areas on what the Department is doing with respect to biosecurity. Most importantly, I want you to know that homeland security is of top priority. It has the personal attention of the Secretary, our subcabinet, agency heads, our USDA employees, and myself.

The Department is a large and complex organization which employs 100,000 people, has offices and installations throughout the world, provides stewardship for 190 million acres of national forest land, and provides more than $100 billion in loans, grants, and services annually. More than one in six Americans participates in programs sponsored by USDA, and many more benefit from the very diverse set of programs the Department operates. In this context, carrying out the Department's responsibilities for homeland security requires a very large effort, and it also requires discipline and focus. Our most intense efforts have, therefore, been directed to seven key areas, which we believe must be addressed if we are to be successful in carrying out our homeland security responsibilities. I would like to give you a brief report on what we are doing in each of these areas.

Protecting U.S. Borders

"USDA has important responsibilities at U.S. borders, airports and ports of entry. The Animal and Plant Health Inspection Service (APHIS) carries out inspections at U.S. ports-of-entry to prevent the introduction of foreign plant and animal pests and diseases, which would be harmful to our country's agriculture. The Food Safety and Inspection Service (FSIS) reviews foreign inspection systems and facilities that export meat and poultry products to the United States and reinspects all imported meat, poultry, and egg products to insure that U.S. requirements are met. Scientific support for these activities is provided by the Department's Agricultural Research Service (ARS). The Office of Inspector General provides audit and enforcement services.

"The Department of Agriculture has been in the business of biosecurity since its inception. As you know, the Department had already been working to strengthen our border inspection systems prior to September 11th due to the presence of foot and mouth disease in the United Kingdom, Europe and South America.

Since September 11th, we have adjusted and strengthened our systems even further. By way of background APHIS, which is in charge of monitoring our borders, has 5,000 inspectors, veterinarians and other personnel helping at 126 ports of entry. In addition, these individuals work with state and industry officials to ensure prevention of harmful animal and plant diseases from entering our country. APHIS has responded by increasing awareness within the veterinarian community. Specifically, the agency has recently conducted an educational teleconference with veterinarian professionals in which diagnostic and foreign animal disease recognition skills were emphasized.

"We are also working closely with our Federal partners including the Federal Emergency Management Agency (FEMA) and the Federal Bureau of Investigation. Our veterinary medical and plant health communities have been put on notice to treat every foreign disease or pest investigation with increased diligence. All APHIS and FSIS field staff has been placed on a heightened state of alert. In addition, the Department is formalizing information flow throughout the our regulated industries to maintain confidence that we are doing everything possible to secure the food supply. Finally, we have established a protocol with the Federal Aviation Administration for the delivery of investigative samples by military transport to our laboratories in the event of another civil aircraft stand down. We must insure the rapid transportation of biological samples to diagnostic laboratories during emergency situations.

Assuring a Safe Food Supply

"For purposes of homeland security, the complete, the complete process of production, processing, storage and distribution of food is important. This includes the seed necessary for production, feed for livestock and poultry, fertilizer for increasing crop yields, and farm equipment and repair parts for the machinery necessary to support agricultural production. Obviously, the protection of the Nation's food supply is a major undertaking and involves the efforts of a variety of USDA agencies and the Department of Health and Human Services through the Food and Drug Administration.

"It is important to realize that the Department of Agriculture has been in the food safety business for almost 100 years since the passage of the original Federal meat inspection legislation in 1906. Over the course of that time, our responsibilities have been expanded and our systems have improved. We have well-established partnerships with other Federal agencies, State and local governments, and with industry. The system was functioning effectively prior to September 11th and is continuing to function effectively.

"The Department's FSIS has fundamental responsibility for meat, poultry, and egg products and carries out its responsibility through a team that includes over 7,600 food inspectors, 200 compliance officers, and 200 laboratory personnel. Since 1996, FSIS has been highly successful in working with industry to install landmark pathogen reduction/hazard analysis and critical control point systems, which greatly strengthened the ability of the inspection system to respond to food safety issues. The FSIS works closely with Centers for Disease Control and Prevention

and state agencies to conduct an ongoing systematic collection of food borne illness data to detect outbreaks and monitor disease trends and patterns. 0

"USDA has other important responsibilities in connection with the food supply: The Commodity Credit Corporation (CCC) because of various activities resulting in the acquisition of commodities; the Food and Nutrition Service because of our efforts to provide food assistance to children and needy families. The Farm Service Agency, because of the ritical linkage provided by that agency to our Nation's farmers; and the Foreign Agricultural Service as the responsibility to gather information on current food and agriculture situations because of the capability of that agency to gather information worldwide. Scientific support for these activities is provided by ARS, and audit and enforcement support is provided by the OIG.

"The Department has taken a variety of actions to further strengthen these systems. USDA has a Food Emergency Rapid Response and Evaluation Team (FERRET), which was authorized by the Agricultural Research, Extension, and Education Reform Act of 1998, and is chaired by our Under Secretary for Food Safety. FERRET is very active in ensuring the necessary USDA-wide coordination of food safety activities. We have put our own personnel on a heightened state of alert. We are working with our cooperators to make sure that they are engaged in a heightened state of alert as well as establishing a Food Biosecurity action team to serve as the arms and legs of our efforts to ensure that we are vigilant in safeguarding foods under USDA's jurisdiction. USDA has been meeting on a regular basis with FDA's food counter terrorism committee. In addition, USDA has recently organized to form the Food Threat Preparedness Network, linking FDA, CDC, FSIS and others to focus on preventative activities to proactively protect our food supply. For instance, the Department of Agriculture is working with industry to develop guidelines for security measures. We continue to provide emergency food relief in support of the Federal Government's efforts in New York.

Protecting and Enhancing Research and Laboratory Facilities

"Science and technical support are a vital component of our overall homeland security efforts. ARS is our principle in-house research agency. APHIS and FSIS also maintain a number of laboratories and methods development centers. In addition, we work closely with our cooperators at 78 land grant universities located throughout the U.S. In short, we have tremendous scientific capability to respond to homeland security issues, but we must maximize security and further improve this capability.

"Since September 11th, USDA has enhanced the security of its research buildings, laboratories, and pathogen inventories, and also established new guidelines for personnel suitability. Those measures include increased USDA security, and additional patrols and surveillance by the Coast Guard of the waters and shipping lanes surrounding our facility at Plum Island, New York. USDA is also making sure that all the work the Department conducts with sensitive materials performed in the most secure locations.

"We have taken two additional actions to provide further assurance that we are doing all of the necessary measures in this area. We have contracted with SANDIA National Laboratory to provide a risk assessment and security analysis of our five Biosecurity Level Three laboratories. The Department has also asked the OIG to conduct reviews of the controls and procedures throughout the Department's laboratory system to ensure that facilities are secure.

Protecting Other Infrastructure

"The Department has a huge infrastructure beyond those particular areas I have already discussed. We have more than 24,000 buildings at 7,000 sites throughout the world. We have responsibility for the National Forest System. The Natural Resources Conservation Service has a variety of responsibilities in rural America, including providing technical assistance to help assess water supply vulnerability.

The Rural Utilities Service (RUS) provides funding for electric, telecommunication, and water and waste disposal systems in rural America. These and other activities are all important in the context of homeland security, and we are doing everything possible to strengthen these activities. For instance, the Forest Service has established additional patrols to improve security on National Forest System lands; RUS is working with its borrowers to improve security where necessary at electric, telecommunications, and water systems financed by the Federal Government.

"At this point, I want to pay particular attention to one of our most important responsibilities – the protection of our own employees. At the USDA headquarters complex, members of the guard force were armed for the first time and will remain armed. Increased numbers of officers have been added to supplement the basic staff. We have technology, within the Department, that enables environmental testing for anthrax. This technology has been used by multiple government agencies during the recent anthrax emergencies. We have used that capability to establish a mobile diagnosis unit at the Washington Navy Yard to furnish rapid responses to possible environmental anthrax detections. This unit is being used for the protection of USDA employees and has also been made available to other Cabinet-level organizations.

"Obviously, the vast majority of the USDA workforce is outside the Washington area. We are working aggressively with all of our agencies to upgrade security wherever necessary for all of our employees. In this regard, one specific action is our effort to expedite and strengthen our system for security clearances. We have hired a contractor to assist in completing the necessary investigations to evaluate the individuals being considered for security clearances.

Securing Information Technology Resources

"In many areas, information technology is at the core of USDA activities. It is used to gather and use crucial information in support of USDA programs. We issue payments to farmers and engage in thousands of other transactions through information technology. We provide the infrastructure that manages the payroll for 500,000 Federal employees, and the Thrift Savings Retirement Plan for all Federal employees. We are vulnerable to security breaches in these areas. "Our Chief Information Officer has overall responsibility for the Department's Cyber Security Program. We are working to strengthen that program through upgraded security policies and standards as well as through increased oversight and guidance for USDA agencies. We have asked all of our information technology processing centers to raise their alert level and insure that system backups are available.

Continuity of Operations

"In February 2001, the Department established an Office of Crisis Planning and Management. The mission of this office is to manage USDA's emergency operations center, coordinate staff from USDA agencies in response to emergencies, provide USDA liaison with the FEMA, and support a variety of other activities necessary to assure the continuation of the Department's operations in an emergency situation. Shortly after September 11th the Office of Crisis Planning and Management began 24 hour a day and seven day a week operations with on-call personnel.
The Department has a detailed Continuity of Operations Plan (COOP) with alternative work sites to enable USDA's leadership to manage essential functions. The Department's COOP plan was implemented in response to the September 11th events, and we are now in the process of using that experience to review and strengthen the plan where such action is necessary.

Audits and Investigations

"As I have mentioned throughout this testimony, we look to our OIG in many areas for audit and investigative support to help us with our homeland security efforts. OIG has focused its efforts on homeland security cases. OIG has accelerated its overall effort to work with USDA agencies in a number of key areas including the security of USDA laboratories, controls over importation of bio-hazardous materials, vulnerabilities in the National Forest System, and cyber security. The work of the

OIG has been very helpful in all of these areas.

"Mr. Chairman, I have tried through this testimony to provide the Committee with a brief report of some of the key activities the Department is carrying out in support of homeland security. As the President has repeatedly stressed, homeland security is a long-term issue. We have a lot more work to do in the Department of Agriculture before we are fully satisfied that we have done everything possible for homeland security.

"However, we need to look at measures to strengthen our already rigorous system of protections. This is particularly true in the area of infrastructure—our research and laboratory capabilities. You have heard the Secretary speak of this issue several times, but we need to ensure investment in the systems that will protect our food system, farmers and ranchers. This takes time and resources; neither of which are unlimited. I will work with Congress in examining these long-term measures to ensure the protection of our farms and food supply.

"We are proud of our employees who provided food assistance in New York and of our Forest Service incident management teams, which provided assistance to the New York Fire Department and FEMA in the immediate aftermath of the September 11th events. We have a tremendous diversity of talent in USDA, and there is no doubt that we will be able to mobilize that talent in support of homeland security.

"One final note I would make has to do with the subject of communication. We simply must do everything possible to communicate to the public the actions we are taking in support of homeland security. The Secretary and other top officials of the Department are issuing public statements and are discussing this topic at every opportunity. USDA's website @ http://www.usda.gov includes access to a series of linkages which contain information about the actions we are taking to keep America's food and agriculture safe. We look forward to a strong and cooperative relationship with this Committee and other Committees in the Congress as we move ahead. I would be glad to respond to your questions or to provide any additional information for the record that you may require

"Again, thank you for the opportunity to talk with you today about this most important issue.

Appendix E

Board on Agriculture and Natural Resources Publications

POLICY AND RESOURCES

Agricultural Biotechnology and the Poor: Proceedings of an International Conference (2000)
Agricultural Biotechnology: Strategies for National Competitiveness (1987)
Agriculture and the Undergraduate: Proceedings (1992)
Agriculture's Role in K-12 Education (1998)
Agriculture's Role in K-12 Education: A Forum on the National Science Education Standards (1998)
Alternative Agriculture (1989)
Animal Biotechnology: Science-Based Concerns (2002)
Brucellosis in the Greater Yellowstone Area (1998)
Colleges of Agriculture at the Land Grant Universities: Public Service and Public Policy (1996)
Colleges of Agriculture at the Land Grant Universities: A Profile (1995)
Designing an Agricultural Genome Program (1998)
Designing Foods: Animal Product Options in the Marketplace (1988)
Ecological Monitoring of Genetically Modified Crops (2001)
Ecologically Based Pest Management: New Solutions for a New Century (1996)
Emerging Animal Diseases - Global Markets, Global Safety: A Workshop Summary (2002)
Ensuring Safe Food: From Production to Consumption (1998)
Exploring Horizons for Domestic Animal Genomics: Workshop Summary (2002)
Forested Landscapes in Perspective: Prospects and Opportunities for Sustainable Management of America's Nonfederal Forests (1997)

Frontiers in Agricultural Research: Food, Health, Environment, and Communities (2002)
Future Role of Pesticides for U.S. Agriculture (2000)
Genetic Engineering of Plants: Agricultural Research Opportunities and Policy Concerns (1984)
Genetically Modified Pest-Protected Plants: Science and Regulation (2000)
Incorporating Science, Economics, and Sociology in Developing Sanitary and Phytosanitary Standards in International Trade: Proceedings of a Conference (2000)
Investing in Research: A Proposal to Strengthen the Agricultural, Food, and Environmental System (1989)
Investing in the National Research Initiative: An Update of the Competitive Grants Program in the U.S. Department of Agriculture (1994)
Managing Global Genetic Resources: Agricultural Crop Issues and Policies (1993)
Managing Global Genetic Resources: Forest Trees (1991)
Managing Global Genetic Resources: Livestock (1993)
Managing Global Genetic Resources: The U.S. National Plant Germplasm System (1991)
National Research Initiative: A Vital Competitive Grants Program in Food, Fiber, and Natural Resources Research (2000)
New Directions for Biosciences Research in Agriculture: High-Reward Opportunities (1985)
Pesticide Resistance: Strategies and Tactics for Management (1986)
Pesticides and Groundwater Quality: Issues and Problems in Four States (1986)
Pesticides in the Diets of Infants and Children (1993)
Precision Agriculture in the 21st Century: Geopspatial and Information Technologies in Crop Management (1997)
Professional Societies and Ecologically Based Pest Management (2000)
Rangeland Health: New Methods to Classify, Inventory, and Monitor Rangelands (1994)
Regulating Pesticides in Food: The Delaney Paradox (1987)
Resource Management (1991)
The Role of Chromium in Animal Nutrition (1997)
The Scientific Basis for Estimating Air Emissions from Animal Feeding Operations: Interim Report (2002)
Soil and Water Quality: An Agenda for Agriculture (1993)
Soil Conservation: Assessing the National Resources Inventory, Volume 1 (1986); Volume 2 (1986)
Standards in International Trade (2000)
Sustainable Agriculture and the Environment in the Humid Tropics (1993)
Sustainable Agriculture Research and Education in the Field: A Proceedings (1991)

Toward Sustainability: A Plan for Collaborative Research on Agriculture and Natural Resource Management
Understanding Agriculture: New Directions for Education (1988)
The Use of Drugs in Food Animals: Benefits and Risks (1999)
Water Transfers in the West: Efficiency, Equity, and the Environment (1992)
Wood in Our Future: The Role of Life Cycle Analysis (1997)

NUTRIENT REQUIREMENTS OF DOMESTIC ANIMALS SERIES AND RELATED TITLES

Building a North American Feed Information System (1995) (available from the Board on Agriculture)
Metabolic Modifiers: Effects on the Nutrient Requirements of Food-Producing Animals (1994)
Nutrient Requirements of Beef Cattle, Seventh Revised Edition, Update (2000)
Nutrient Requirements of Cats, Revised Edition (1986)
Nutrient Requirements of Dairy Cattle, Seventh Revised Edition (2001)
Nutrient Requirements of Dogs, Revised Edition (1985)
Nutrient Requirements of Fish (1993)
Nutrient Requirements of Horses, Fifth Revised Edition (1989)
Nutrient Requirements of Laboratory Animals, Fourth Revised Edition (1995)
Nutrient Requirements of Nonhuman Primates, Second Revised Edition (2003)
Nutrient Requirements of Poultry, Ninth Revised Edition (1994)
Nutrient Requirements of Sheep, Sixth Revised Edition (1985)
Nutrient Requirements of Swine, Tenth Revised Edition (1998)
Predicting Feed Intake of Food-Producing Animals (1986)
Role of Chromium in Animal Nutrition (1997)
Ruminant Nitrogen Uses (1985)
The Scientific Basis for Estimating Air Emissions from Animal Feeding Operations: Interim Report (2002)
Scientific Advances in Animal Nutrition: Promise for the New Century (2001)
Vitamin Tolerance of Animals (1987)

Further information, additional titles (prior to 1984), and prices are available from the National Academies Press, 500 5th Street, NW, Washington, D.C. 20001, 202-334-3313 (information only). To order any of the titles you see above, visit the National Academy Press bookstore at http://www.nap.edu/bookstore.

About the Authors

COMMITTEE ON BIOLOGICAL THREATS TO AGRICULTURAL PLANTS AND ANIMALS

Harley W. Moon, D.V.M., Ph.D. (NAS) (Chair), is F. K. Ramsey Chair of Veterinary Medicine at Iowa State University. He has been a member of the National Academy of Sciences since 1991. He has served on several NRC committees including the Committee on Ensuring Safe Food from Production to Consumption and the Committee on Eradicating Bovine Tuberculosis. He is currently chairperson of the Board on Agriculture and Natural Resources. Dr. Moon is most widely recognized for his contributions to the basic understanding of intestinal diseases of humans and animals. His expertise includes the development of vaccines for preventing *E. coli* infection in farm animals, livestock disease eradication, infectious diseases affecting humans and animals, and prevention of edema disease in swine with genetically modified vaccines. Dr. Moon has served on numerous advisory committees, including the World Health Organization's Expert Panel on Enteropathogenic *E. coli* and Working Group on Immunology and Vaccine Development for Bacterial Enteric Infections, the Department of Agriculture's Task Force on Scrapie and Bovine Spongioform Encephalopathy, Pioneer Hi-Bred International's Institutional Biosafety Committee, and Council for Agricultural Science and Technology Task Force on Antibiotics in Animal Feeds. He presently serves as a consultant to Agricultural Technology Partners, LP, and owns and manages a farm in Iowa. His scientific publications number in excess of 200, with numerous book chapters on aspects of infectious disease. Before his current position, Dr. Moon was director of the Plum Island and National Animal Disease Centers, ARS/USDA, and professor in the Department of Veterinary Pathology at Ohio State University. He received his B.S., D.V.M. and Ph.D. at the University of Minnesota.

Michael S. Ascher, M.D., FACP, is Chief of the Viral and Rickettsial Disease Laboratory Branch, Division of Communicable Disease Control, at the California Department of Health Services. Since the fall of 2001, he is on temporary assignment to the Office of Public Health Preparedness, Immediate Office of the Secretary, U.S. Department of Health and Human Services in Washington as the Associate Director for laboratory and technical issues. Dr. Ascher, a native of Illinois, graduated from Dartmouth and Harvard Medical Schools. He trained in Internal Medicine, Infectious Disease and Immunology at Bellevue Hospital Center in New York City. He served in the U.S. Army as Chief of Medicine and in the Bacteriology Division at USAMRIID and as a traveling fellow at the Royal College of Surgeons of England. He joined the Division of Infectious Disease at the U.C. Irvine College of Medicine in 1978. In 1985, he moved to Berkeley as a Public Health Medical Officer in the Viral and Rickettsial Disease Laboratory of the California Department of Health Services and was appointed Chief of the Laboratory in 1995. He is a Lecturer in the School of Public Health of the University of California, Berkeley. In the area of biological defense, he has served on the Armed Forces Epidemiological Board and an interagency advisory panel on Biological Warfare Preparedness for the 21st Century and currently consults in this area of biological defense preparedness to the Department of Defense, Centers for Disease Control and Prevention, Mitre Corporation, the National Domestic Preparedness Office of the FBI, the Association of Public Health Laboratories, and national laboratories. He is a founding member of the Working Group on Civilian Biodefense of the Center for Civilian Biodefense Studies of the Johns Hopkins University School of Public Health. Prior to his new assignment, he was the lead for biological defense activities in the California Department of Health Services and Principal Investigator of the CDC grant to the State for preparedness and response. Dr. Ascher's research interests include mechanisms of protective immunogenicity of microbial vaccines, advanced methods for diagnosis of infectious diseases and fundamental issues of HIV pathogenesis. He is a member of numerous scientific societies and has over 90 publications.

R. James Cook, Ph.D. (NAS), is the Endowed Chair in Wheat Research at Washington State University, Pullman, WA, a position assumed April 1, 1998. Prior to this appointment, he worked as a Research Plant Pathologist with USDA-ARS at Pullman from 1965 through March of 1998, conducting research on biological approaches to control root diseases of Pacific Northwest wheat. He completed his B.S. in 1958 and M.S. in 1961, both at North Dakota University, and his Ph.D. in 1964 at the University of California, Berkeley. He worked as a NATO Postdoctoral Fellow to Australia in 1964-65, a Guggenheim Fellow to Australia in 1973-74, Visiting Scientist to Cambridge, England in 1981, and as a Fellow of the Japanese Society for Promotion of Science in Japan in 1983. He has served as President of the American Phytopathological Society, and from 1988-1993 as President of the International Society of Plant Pathology. He has co-authored two books on biological control of plant pathogens and one book on wheat health management. He is Fellow of the American Phytopathological Society and the American Association for the Advancement of Science and was elected to the National Academy of Sciences in 1993.

David R. Franz, D.V.M., Ph.D., is currently Vice President of Chemical & Biological Defense Division at Southern Research Institute. He has served in the U.S. Army Medical Research and Materiel Command for 23 of his 27 years on active duty. Dr. Franz has served as both Deputy Commander and then Commander of the U.S. Army Medical Research Institute of Infectious Diseases (USAMRIID) and as Deputy Commander of the U.S. Army Medical Research and Materiel Command. Prior to joining the Command, he served as Group Veterinarian for the 10th Special Forces Group (Airborne). Dr. Franz served as Chief Inspector on three United Nations Special Commission biological warfare inspection missions to Iraq, and as technical advisor on long-term monitoring. He also served as a member of the first two US/UK teams that visited Russia in support of the Trilateral Joint Statement on Biological Weapons, and as a member of the Trilateral Experts' Committee for biological weapons negotiations. While at the Medical Research and Materiel Command, he was assigned to four of its laboratories, personally conducting research and publishing in the areas of frostbite pathogenesis, organophosphate chemical warfare agent effects on pulmonary and upper airways function, the role of cell-mediated small vessel dysfunction in cerebral malaria, and medical counter-measures to biological agents. Dr. Franz was Technical Editor for the Textbook of Military Medicine on Chemical and Biological Defense released in 1997. He has been an invited speaker at many nationally and internationally recognized organizations. Dr. Franz holds a D.V.M. from Kansas State University and a Ph.D. in Physiology from Baylor College of Medicine.

Marjorie Hoy, Ph.D., is the Davis, Fischer and Eckes Professor of Biological Control in the Entomology and Nematology Department of the University of Florida, Gainesville. Her research involves biological control of pests of citrus, developing methods for genetic improvement of arthropod natural enemies, developing methods for monitoring biotypes used in classical biological control programs and for monitoring genetic quality of natural enemies during rearing. Dr. Hoy is author of a textbook, Insect Molecular Genetics, which reviews the potential applications of molecular genetic techniques for solving problems of interest to entomologists. She was also involved in the organization of a workshop to evaluate "Potential Risks Associated with Releases of Transgenic Arthropod Natural Enemies." She currently serves on the USDA Advisory Committee on Agricultural Biotechnology. Dr. Hoy earned her MS (1966) and Ph.D. (1972) in entomology from University of California, Berkeley.

Donald F. Husnik, B.S., has been a consultant on national and international phytosanitary issues since leaving USDA in 1996. He has consulted with organizations and governments on plant quarantine systems pertaining to the import and export of agricultural products. He has 35 years of experience in designing, directing, and implementing national and international programs to protect the health of plant resources from destructive pests and diseases. From 1995-1996, he served as the Deputy Administrator of Plant Protection and Quarantine, Animal and Plant Health Inspection Service, USDA. He provided leadership to U.S. agricultural pest exclusion, detection and surveillance, emergency response and pest management programs; including the Agricultural Quarantine Inspection Program, Cooperative Agricultural Pest Survey Programs, and a number of cooperative federal/state/grower exotic pest eradication and control programs. He was the principle APHIS representative to the National Plant Board and the North American Plant Health Organization. From 1962 to 1995 he served APHIS in various technical and managerial capacities. He chaired a multidisciplinary team that designed the APHIS Center for Plant Health Science and Technology. From 1988-1991 he served as Director, Policy and Program Development, APHIS. He directed the work of senior policy analysts, planning and evaluation specialists, regulatory analysts and risk assessment specialists. He led APHIS efforts to develop internationally accepted risk assessment models. He was a member of the U.S. Senior Executive Service from 1981 to 1996. Mr. Husnik received his B.S. degree in Agronomy from the University of Minnesota.

Helen H. Jensen, Ph.D., is Professor of Economics and Head of the Center for Agricultural and Rural Development's Food and Nutrition Policy Division at Iowa State University. Her current research focuses on food programs and policies, including food safety. Dr. Jensen's major areas of research are food demand analysis, food assistance and nutrition policies, issues related to food security, and the economics of food safety and food hazard control options. Dr. Jensen currently serves on the editorial board of the journal of *Agricultural Economics* and has been an active member of the American Agricultural Economics Association where she has chaired several working committees. She was a member of the National Research Council's Panel on Animal Health and Veterinary Medicine from 1995-1998. She joined the faculty at Iowa State in 1985 and holds a Ph.D. degree in Agricultural Economics from the University of Wisconsin.

Kenneth H. Keller, Ph.D. (NAE), is Charles M. Denny, Jr. Professor of Science, Technology and Public Policy in the Hubert H. Humphrey Institute of Public Affairs at the University of Minnesota. Educated at Columbia and Johns Hopkins, he spent most of his career at Minnesota where he joined the faculty of Chemical

Engineering and Materials Science in 1964, became Vice President for Academic Affairs in 1980 and President of the University in 1985. From 1990 to 1996, he was Senior Fellow for Science and Technology at the Council on Foreign Relations and, for two of those years, Senior Vice President for Programs. His research interests include the impact of science and technology on international politics and economics, the policy issues raised by high technology medicine, and the role of American institutions of higher education in research and development. Dr. Keller has served as a member of the Commission on Physical Sciences, Mathematics and Applications of the National Research Council, the Science and Technology Advisory Panel of the National Intelligence Council, and the Advisory Board of the Institute for Education and Training of the RAND Corporation. He was recently elected to the National Academy of Engineering. Dr. Keller also chairs the Medical Technology Leadership Forum, is Vice President of the American Institute of Medical and Biological Engineering, and is on the board of trustees of LASPAU: Academic and Professional Programs for the Americas, and of the Science Museum of Minnesota. He earned master's and doctorate degrees in chemical engineering from Johns Hopkins University.

Joshua Lederberg, Ph.D. (NAS, IOM), is a geneticist and microbiologist who received the Nobel Prize in 1958 for his work in genetic structure and function in microorganisms. Dr. Lederberg has served as professor of genetics at the University of Wisconsin, then at Stanford School of Medicine, and at the Rockefeller University since 1978. From 1978 to 1990, he served as president of the Rockefeller University. He has been a consultant on health-related matters for government and the international community, having had long service on WHO's Advisory Health Research Council. He received the US National Medal of Science in 1989, where his consultative role was specifically cited. He has served as Chairman of the President's Cancer Panel, and of the Congress' Technology Assessment Advisory Council, as well as on several national and international panels addressing the threat of bioterrorism. Dr. Lederberg has spoken nationally and internationally on bioterrorism, notably at a 1999 NAS gathering during which he briefed President Clinton and his top security advisers. Dr. Lederberg recently published a book on the subject, Biological Weapons: Limiting the Threat. He is currently Sackler Foundation scholar and professor-emeritus of molecular genetics and informatics. Dr. Lederberg continues his research activities at Rockefeller in the field of interactions of gene functionality and mutagenesis in bacteria. Dr. Lederberg was elected to the National Academy of Sciences in 1957 and to the Institute of Medicine in 1971 and is a foreign member of the Royal Society. He received his Ph.D. in Microbiology from Yale University in 1947.

Laurence V. Madden, Ph.D., is Professor of Plant Pathology at Ohio Agricultural Research and Development Center and Ohio State University in Wooster.

He studies the epidemiology of plant diseases with a special emphasis on the influence of the biotic and abiotic environment on epidemics. Dr. Madden's approach is quantitative in nature and models are employed to describe disease development. He has presented at several conferences on the subject of agricultural bioterrorism. At the 1999 annual meeting of the American Phytopathological Society, Dr. Madden unveiled a general probabilistic model for assessing the risk of crop bioterrorism under various scenarios. His model depends upon the probabilities of pathogen introduction, initial establishment, disease spread, and successful confinement or control. Dr. Madden's research is conducted on several diseases with major emphasis on virus and phytoplasma diseases of maize and fungal diseases of fruit crops; a large focus has been to elucidate the effects of rainfall on dispersal of plant pathogens. The general goals of his research program include: development of microprocessor-controlled forecasting systems to predict disease outbreaks and schedule efficient fungicide applications; development of fundamental models for understanding pathogen/vector relations and disease/microclimate interactions; characterizing the relationship(s) between disease intensities and crop losses; improving analytical techniques for analyzing and comparing epidemics; and understanding the interaction of vector population dynamics and epidemic development. Dr. Madden earned his Ph.D. in Plant Pathology in 1980 from Penn State University. He is a Fellow of the American Phytopathological Society, American Association for the Advancement of Science, and the Linnean Society of London. He is a former president of the American Phytopathological Society.

Linda S. Powers, Ph.D., is Director of the National Center for the Design of Molecular Function, Professor of Electrical and Computer Engineering, Professor of Biological and Irrigation Engineering, and Adjunct Professor of Physics at Utah State University. Before joining the USU faculty in 1988, Dr. Powers was a member of technical staff of AT&T Bell Laboratories for 13 years. From 1978-1998, she was an Adjunct Professor of Biochemistry and Biophysics at the University of Pennsylvania Medical School and she has also been a Visiting Fellow in the Department of Chemistry at Princeton University. Dr. Powers has a broad scope of expertise from biochemistry to electrical engineering, and has considerable experience in heme protein catalysis, structural biology, and the design and construction of optical and X-ray instrumentation. Her current research areas include detection of viable microbes on surfaces and the development of microbe detection technology. Dr. Powers was a pioneer in the use of X-ray absorption spectroscopy for the investigation of biological problems and has authored more than 100 technical publications in refereed journals and books. She has served on several advisory boards for the American Physical Society, on editorial boards of *Biophysical Journal* and *International Series in Basic and Applied Biological Physics,* and recently on the NRC committee that authored the 1999 report *Chemical and Biological Terrorism: Research and Development to Improve Civilian*

Medical Response. Dr. Powers is a Fellow of the American Physical Society (1983) and the American Institute of Chemists (1987) and her honors include the US Bioenergetics Award of the Biophysical Society (1982), the Outstanding Researcher of the Year Award from Utah State University (1994), and the State of Utah Governor's Medal for Science and Technology (1994). Dr. Powers completed her M.A. in Physics and Ph.D. in Biophysics (1976) at Harvard University.

Alfred D. Steinberg, M.D., is a consulting physician-scientist for the MITRE Corporation, a non-profit organization, where he provides advice to several government agencies on molecular genetics, immunology, microbiology and issues associated with bioterrorism. He has broad knowledge of the range of issues associated with biological threats to plants, animals, and humans. He currently serves as executive secretary for several MITRE committees and is involved in activities focusing on bioterrorism and remote sensing of human infectious diseases and animal/plant pathogens. Dr. Steinberg is now serving as a member of the National Intelligence Council's 2015 Threat Panel. He was Chief of the Cellular Immunology Section at NIH from 1981 to 1992 and served on several NIH committees. He is a licensed physician in the states of New York, Maryland, and California, and is board certified in rheumatology and internal medicine. Dr. Steinberg has published over 475 papers, many in the area of developing innovative clinical treatments for immune diseases, and he was listed among the 300 most quoted authors in the world from 1973 to 1984. He has served on the editorial board for the *Journal of Immunology*, *Journal of Immunopharmacology*, *African Journal of Clinical Immunology*, *Clinical and Experimental Rheumatology*, *Journal of Autoimmunity*, and *Acta Pathologica, Microbiologica, et Immunologica Scandinavica*. Dr. Steinberg earned his M.D. from Harvard Medical School in 1966.

Consultant

Al Strating, D.V.M., M.S., is a collaborating consultant for Animal Health Solutions International, LLC. He is a veterinarian with post-doctoral work in microbiology, and has dedicated much of his career to national and international management and control of major livestock diseases while working for the USDA Animal and Plant Health Inspection Service (APHIS). Positions with APHIS included Associate Administrator, Director of the APHIS Centers for Epidemiology and Animal Health, Regional Director, and Director of the National Veterinary Services Laboratories. He served for 3 years as vice chairman of the Standards Commission of the Office of International Epizootics in Paris, France. He has also worked as a veterinary practitioner, and as a consultant for USDA, US Agency for International Development, and National Association of State Departments of Agriculture. He is a member of the American Veterinary Medical Association, the US Animal Health Association, and the American Association of Veterinary Laboratory Diagnosticians. Current areas of study include heartwater disease in Africa and the Caribbean, and evaluation of the adequacy of USDA measures for keeping the U. S. free of exotic animal diseases. Dr. Strating has published over 30 scientific articles in the fields of microbiology, immunology, epidemiology, and food safety. Dr. Strating earned his D.V.M from the University of Minnesota in 1965 and his M.S. in Microbiology from Colorado State University in 1979.

Liaison, Committee on Agricultural Biotechnology, Health, and the Environment

Robert E. Smith, Ph.D., is president of R. E. Smith Consulting, Inc. in Newport, Vermont. Dr. Smith has an extensive background and perspective regarding food technologies, nutrition, and research and development. His prior positions include president of the Institute of Food Technologists, senior vice president for corporate research at Nabisco, Inc., and senior vice president of research and development at Del Monte Corporation. His research focuses on the nutritional quality of human and pet foods, particularly protein and amino acid requirements and interrelationships. Dr. Smith received his Ph.D. degree in animal science, with an emphasis on nutrition and biochemistry, from the University of Illinois. He completed his M.S. and B.S. degrees in animal and poultry nutrition at McGill University.